RASMUS

A TELEVISION TALE

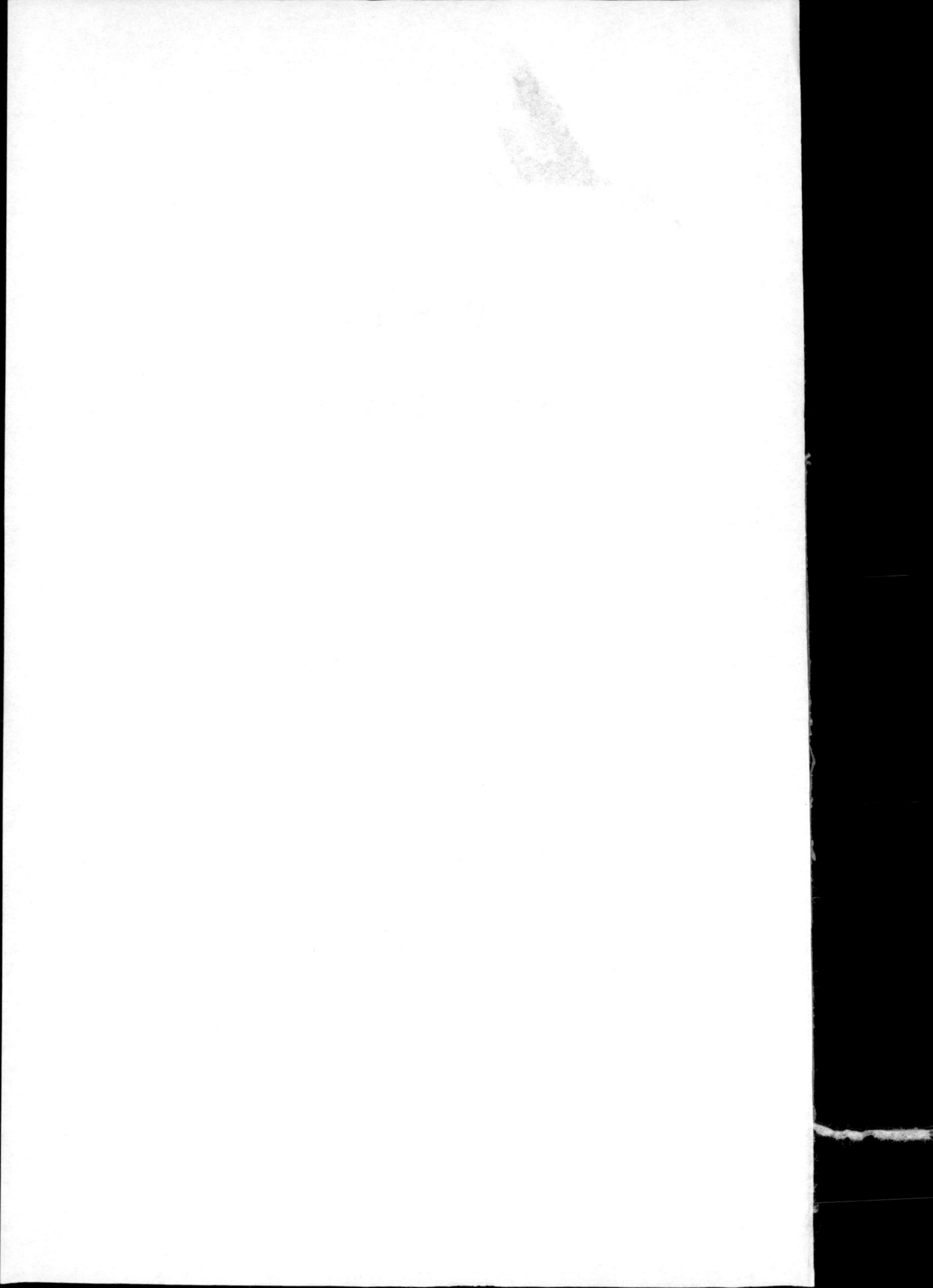

RASMUS

A TELEVISION TALE

P.J. Vanston

Matador
9 Priory Business Park
Wistow Road
Kibworth Beauchamp
Leicester LE8 0RX, UK
Tel: 0116 279 2299
Email: books@troubador.co.uk
Web: www.troubador.co.uk/matador

ISBN 978 1785893 612

British Library Cataloguing in Publication Data.
A catalogue record for this book is available from the British Library.

Printed and bound in the UK by TJ International, Padstow, Cornwall
Typeset by Troubador Publishing Ltd, Leicester, UK

Matador is an imprint of Troubador Publishing Ltd

About the Author

PJ Vanston is the author of several books including the darkly satirical *Crump* (Matador, 2010), as well as several prize-winning short stories.

Writing as Jem Vanston, he is the author of the much-loved *A Cat Called Dog* (2013, 2015) and its sequel *A Cat Called Dog 2* (2016), and is working on further children's books.

A former teacher/lecturer and freelance journalist, he now runs his own editing and proof reading agency.

A published songwriter, he was responsible for writing and recording (as part of *Little Red Dragons*) the charity football song *Come on Wales* for Euro 2016. His songs have recently been made available to established recording artists.

Born and brought up in Dartford, Kent, the author now lives in Swansea.

Author website: www.vanston.co.uk

Twitter: @acatcalleddog

Begin at the beginning and go on till you come to the end: then stop.
Lewis Carroll. Alice's Adventures in Wonderland.

Throughout the centuries there were men who took first steps down new roads armed with nothing but their own vision.
Ayn Rand. The Fountainhead.

Reality leaves a lot to the imagination.
John Lennon

The definition of freedom is being fearless.
Nina Simone

Prologue

The world is television.

The sky flickers blue electric dark. Streets are silent empty. Blue-white bursts of lightning flash like paparazzi on the skin of the night.

A deep thud of thunder sucks the silence from the air. It groans and roars – yet, all around, the city does not stir. It is almost 4am, and London sleeps. Its tangle of streets glow sodium orange, glistening rain-pocked roads reflecting the white light of empty offices for no-one to see. Nothing moves under the storm – no cars, no people, no life. Nothing.

And through it all, between banks crammed with concrete and glass, flows the river, broad and black and old. It sweeps from west to east, from leafy suburbs, past the palaces and bedsits of Kensington, and on through the heart of London – past Westminster and St Paul's, the old stones of history and heresy alike – before eeling out to the estuary to be swallowed whole by the sea. It bubbles under bridges it will one day mock and destroy, seemingly not a single body of water, but a seething mass of liquid individuals, each with a character of its own – brash and bold here, skittish and nervous there – but always the water as thick and black as old, cold blood. The river has been here longer than all of this – this little city of stone and glass – and will remain long after its buildings have been ground to sand and dust too.

Something stirs below the heavy stain of sky. An office block – a taller tower than the others – in Hammersmith. Fourteen floors of black glass and steel. Something on the roof, by the air-conditioning vents breathing steam. It is a man, hunched and head held low to the lashing rain. He is shoeless, and the skin of a T-shirt clings to him. The man walks unevenly, staggers, then throws back his head: a last glug from the glint of a bottle, which he then hurls high into the air, its trajectory destined to shatter in some empty street below. He staggers again - a clumsy puppet dance in the rain - almost falls, then rights himself and walks on until he is standing at the edge of the office block roof. The sky groans and growls, the storm lashing harder now, as if insulted by this tiny human presence beneath its massiveness and might.

The man steps unsteadily up onto the small wall that borders the roof, holds tight to a railing, and lifts his head to the sky. He seems to be talking - shouting, or perhaps laughing. The sky swirls angrily in response, a looming monstrous vortex above him. And then the man looks down - down at the street below, its mosaic of tiny paving slabs and miniature cars parked along it. Down there, in that finality of forever, lay peace - there was nothing to cause him pain down there. He was just one bloody splinter of bone away from the perfection of nothingness - just one step, off that roof, would launch his first and final flight through the rain to his eternal rest.

Just then, the sky rumbles. A voice. It sounds like a voice, anyway. Deep and dark and ancient. The man looks up into the eye of the storm. Eternity stares back.

A bolt of lightning strikes him, connects ground to sky in a bright blue final flash.

The world is television. And it is all over.

But then, it has only just begun.

PART ONE

THE BEGINNING

Execution Night – Live

Two men, sitting on the stone floor of a prison cell.

One man is smaller than the other – fatter, with an unshaven cherub face. His eyes are closed, and he is mumbling to himself. Prayers, probably. The other man watches him without expression. This man is thin and drawn, and looks tall. Both men look exhausted and drained, like corpses. They each know one of them will be dead within the hour, because that is the will of Allah – or, at least, the television company.

The thin man just knows it will be him and is resigned to his fate, so says nothing. He has spoken very little during their time in the cell. He hasn't tried to play up to the always-on camera, but has tried to ignore it, as he has ignored the buzzing flies and the stench of death all around. Mostly, he has said nothing. Except with his eyes: they say all he wants or needs to say.

The little fat man, Ali, has been much more entertaining, and is much better TV. He has played to the camera, begged for mercy in comically-broken English, and talked of how he is a loving and dutiful husband to his wife and a devoted father to his seven children, whom he names one by one and describes fondly several times every day. TV screens all over the world have sweated his misery to audiences far and wide. He has explained how it has all been a big mistake – how he ended up killing that girl. He thought she was a criminal, out to rob him, to attack him – that is why he hurt her. He had never been unfaithful to his wife – not once. He loved her. He *loves* her. He would never want to rape a young girl like that – a girl the same age as one of his daughters – and he would never kill her even if he did rape her. How could they think such a thing? It had all been a terrible misunderstanding.

The tall thin man is called Mohammed. He watches his cell-mate when he is speaking, but has never replied to any of Ali's attempts to start conversation. Why he agreed to all this in the first place he does not really know. He is not afraid of death, and sometimes wishes he'd been executed as planned after his trial, and wasn't now sitting in a stinking prison cell, waiting for the end. Maybe he deserves to die. He did, after all, rape a teenage girl before cutting her throat so she wouldn't talk – (and she was not the first, by any means) – and he was chased and caught straight afterwards, with her blood still wet like sweat on his hands. Just his bad luck. He cannot deny what he did, and never would. He did it – yes, and he enjoyed it too. Allah in his infinite and unknowable wisdom has made him this way, just like his father – the man who used to beat and rape him – whom he saw beheaded when a child. Mohammed will welcome death as a friend whenever it comes.

Ali stops mumbling and opens his eyes. Mohammed stares at him, blankly. Ali starts to sob like a child. The cameras and microphone are set up outside the bars of their cell – the bars take up one whole wall of it, so there is no privacy. The all-seeing eye of the lens misses nothing – it watches every sob and wail, zooms in on every glistening bead of sweat on skin. That is what it is there for – to capture reality as it is. Mohammed does not know how Ali can have such disrespect for himself. What kind of a man is he? But then, Mohammed knows that he himself also agreed to participate in this game – this unholy infidel game show – just for the chance to keep breathing and to live out his life. In truth, courage is an illusion. Every man is a coward in the end.

But Mohammed now knows that there is far more to fear from the camera than from death itself: that evil unblinking eye will steal your soul and eat into your heart until it stops living, even though it will still beat. It will kill you and leave you alive, but with a dead soul. That is far, far worse than a quick slice of the knife by a skilful executioner which, when done well, is the manner of execution to be preferred. Mohammed remembers that his father was smiling when his head was severed, and his eyes still blinked and winked at him when held up high by the executioner to cheers from the crowd. He always did like being the centre of attention.

The guards unlock the doors and lead the prisoners out to a holding room. On the wall, a giant wide-screen TV, like the eye of God. Mohammed and Ali watch it, transfixed.

*

A blast of thumping hardcore music.

A spinning camera circling 365 degrees.

A manic cheer shrieks and screams into the microphones.

'Execution Night – Live' has begun.

'Hello London!' yells Alicia McVicar through the noise, 'And...hello world!' She pauses until the shrieks and whoops of the crowd acknowledge her presence.

She smiles confidently, waves at the crowd and the cameras, jumps up and down a couple of times to make her breasts bounce, then takes a deep breath.

'Bring it on!' she screams, and the crowd yells back its approval, a willing congregation yearning for the enlightenment of reality TV.

Alicia always says 'bring it on' with a mock-male-aggressive sneer on her face when she does so – it is a trick she has picked up from other female TV presenters. Producers say that audiences like it when young attractive women are aggressively full-on and in-yer-face like this. It makes them look strong, confident and in control, so turns on the boys as well as giving the girls a role model to identify with – in such a way are great ratings achieved. These words are her trademark

– her catchphrase – part of her TV identity and so part of her reality. Alicia screams the words again:

'Bring it on!'

'Whoooooaaaaaahhhhhh!' screams the crowd, whooping and hollering hysterically.

Cheers cheers cheers.

'Welcome...everyone...to X-TV...and Execution Night – Live...coming at you live and direct from London town! Bring. It. On!!! Yay!!!'

Alicia grins and waggles her chest at the camera which is there to capture just such close-up presenter shots. She specifically chose the skimpiest dress she could find to wear, and one that would mean she would be freezing cold on that chilly English April evening. That way, even if this particular TV show bombs, other television producers and executives who may be watching won't be able to get the sight of her erect nipples out of their heads, and she'll be a shoo-in for other TV presenting work.

The crowd roars like escaped mental patients from some Hammer Horror Victorian asylum, or perhaps like disturbed apes in a zoo, hungry for the attention offered by a camera lens: a peculiar sort of behaviour that one can see nowhere else but on TV. The warm-up acts, the music, the time, the place – all have come together to whip the audience into a frenzy. Most are young, but some over-thirties are squeezed in amongst those taut teenage bodies. A majority of the crowd is female, which is not unusual for reality TV. They are the baying mob, the tattooed tricoteuses of a new revolutionary age. And they all look manic, though whether that's a natural high or due to narcotic assistance is unclear.

What is clear is that they are all desperate to see if their man – the man they voted to spare – is, indeed, spared; and, more importantly, to watch the loser have his head cut off, which is what the viewers at home are all really voting for. Because one of the men whose faces peer out from the giant screen at the warehouse yard in East London will soon be put to death, just like men have been publicly slaughtered for the entertainment of spectators in all cultures for thousands of years. One could even call 'Execution Night - Live' a return to traditional, old-fashioned values, in its way. But it is also a TV first, and everyone present knows they are part of history. There is an electricity - a buzz - a weird alchemy in the air that is only really there very occasionally on TV, no matter how hard producers and directors try to artificially create it.

A digital clock next to the enormous screen counts down the seconds - there are two minutes left. Less.

It has not gone unnoticed by the media in the UK and internationally that X-TV's initial live real-time ratings for 'Execution Night - Live' are high, and rising. They are beating all other TV channels hands down too - though the full ratings figures will only be available after some days. Simultaneously, all over the world, thousands of producers are kicking themselves they didn't think of putting live executions on TV first, and wishing they worked for X-TV.

Rasmus watches all this on the widescreen plasma TV in the living room of his Holland Park mansion. He sips tea as he watches the evening's events unfold. His assistant, Thursday, stands and watches too, though he has never been a fan of TV really, not even the violent shows, and only rarely watches that stuff on the internet - real violence happening at the end of your fingertips is much more exciting.

At the BBC, most senior managers are watching closely - quite a number have voted in the X-TV execution poll, though they would never admit to it.

Politicians watch too, unsure if all this is legal, but powerless to stop it, even if some want to. Many MPs have voted in the X-TV poll, feeling somehow involved in these two men's fate, and are as glued to their screens as much as the general population. Not even the MPs who deplore the broadcast of an execution and the reduction of a man's last hours to entertainment can deny the power of shows like this, despite the unpleasantness, and envy the popularity of reality TV and the way it seems to inspire the young to vote.

Everybody, it seems, is now watching X-TV. In every street and lane and road and avenue and cul-de-sac, in flats and semis and bungalows and mansions and terraces, from the north of Scotland to the west of Wales to the south of England, the whole nation is watching – in their millions. The internet's reach means that many millions more are also watching all over the world, despite no promotion of the show outside Britain and the time difference. The ratings are particularly high in the Muslim world, Israel and the USA.

And then the time comes. The clock ticks down to zero, and voting has ended.

Soon, the show will reach its climax. The people out there have waited a week for this moment, the tension increasing day by day and hour by hour – and now, second by second. The viewers have been told that the vote has been close all week, but they have no idea how close. Electric tension fizzes and spits in the air.

Suddenly, the thumping music stops dead. Alicia peers up at the huge screen at the outdoor broadcast in East London and pauses – just enough – to allow the whispers to die down: the tension builds as intended.

'And here they are – Ali and Mohammed – in Afghanistan,' she says, as if introducing child contestants in a fancy dress competition.

More cheers and shrieks and whooping: everyone sounds like American white trash on TV shows these days.

Alicia is still not sure exactly where Afghanistan is – although she did know people who had an Afghan hound once, so they've got nice dogs over there, at least. But she also knows that this lack of knowledge doesn't matter: she is paid to present TV shows, not to know stuff. How many people in the audience know where that country is or anything about it? And anyway, who cares? Ignorance is cool for the internet generation, she knows. Why waste time learning any fact about anything when you can just Google it and find the answer in two seconds flat anyway?

The noise dies down and the steady sound of a clock ticking is relayed to the audience and the viewers at home. Of course, it's not the sound of the real clock, but a sound effect synchronised precisely with the digital clock everyone is looking at. Even the time is fake on TV: an illusion like everything else.

Alicia puts on her 'worried-and-concerned' look – the one she spent ages practising in the mirror, together with the pouty prick-teaser look, the poor innocent vulnerable victim look, and the 'I'm-about-to-give-you-the-blowjob-of-your-life' look – all of which have proved very useful indeed in her TV presenter career.

'One of these two men... will soon... be dead...' she says, breathlessly, in a sultry low whisper.

Gasps.

'His head... will be cut from his body...'

Gasps. Pause.

'By an executioner... with a great big sharp knife...'

Gasps. Pause. Wait for it, wait for it.

'But the other man – the one you, the X-TV viewers, voted for...'

Wait. Wait. Wait. Then she shouts it out loud:

'Will be set freeeeeeeee!'

Cheers and roars from the hyped-up psyched-up crazy crowd.

'Isn't that amaaaaaazing?! Bring it on!'

Cheers and screams and shrieks – yelling and hollering and whooping – a last release of pent-up tension before the inevitable.

In that moment, Alicia sincerely believes that she is feeling what Nelson Mandela felt when he was freed from decades in captivity – or, at least, what the TV presenter who announced it was feeling. Alicia has watched the archive footage online many times. It's all just *so amazing!*

The audience quietens down as they watch – on the huge screen – the men, Ali and Mohammed, with hands tied behind their backs, being led from the holding room by masked guards and taken, with a hand-held camera following, into the courtyard outside.

And there he is – the executioner – a black leather mask for a face. He is a big – no, a huge – man, who probably weighs as much as the two prisoners together. And in his enormous hands, each the size of a plate, sits a large, curved knife. The condemned men glimpse the steel glint of the blade and tense – their weary state has now become one of nervous vivid wakefulness. Ali feels shaky and sick, and swallows back down some acid vomit that has leaked into his throat – if only he could have a drink of water!

They are then led to the place where it will happen – in front of the masked executioner. Ali is mumbling prayers. Mohammed is resigned to his fate and is silent, breathing each breath so as to taste its sweetness more.

One of the guards touches the prisoners' shoulders – a sign to kneel. They do so. Cold sweat glistens on their bare necks as they sense the shadow of the executioner behind them.

The crowd in London is quiet now – hushed, like at a World Cup penalty shoot-out. It looks at the screen, each individual a part of a whole, each a small cell in a huge body of humanity, thinking and reacting as one enormous organism, a Leviathan lusting for blood.

The scene-setting ominous music has stopped. Instead there is a hush – a tense tingling hush in the air. Alicia McVicar is in full control. She is good at this and she knows it.

'The voting...has finished...The votes...have been counted. So now...'

Hush.

'I can finally reveal.'

Hush.

'That the man...whose life... will be spared...tonight...is...'

Hush and held breath.

'By a margin... of 51.2% to 48.8%...'

Gasps in the crowd at the closeness of the vote.

C'mon c'mon c'mon, tell us!

Silence.

Hush. Hush! HUSH!

TELL US!

'Is...'

JUST FUCKING TELL US!!!

'Mohammed!'

Cheers, screams, wails, gasps, yells. In London. And all over the UK and the world.

At almost the exact same moment that Alicia McVicar says those words, the executioner's razor-sharp knife slices into Ali's neck, cuts his throat deep in one swift movement. The blade goes in easily, as if it were slicing cheap cheese.

Ali looks startled. The eyes widen in his face as the blade cuts into his double chin. He gurgles and tries to scream, but the sound simply bubbles in the blood that is now gushing from the stump of his neck like a little red fountain. It has taken only three cuts to remove his small, round head from his body. Cutting edge TV...

Meanwhile, Mohammed's face is expressionless, despite his amazement at his own uncut throat. He is surprised that he can still taste his breath, sense his heart pumping in his chest, feel himself alive and intact.

Ali, by contrast, can taste nothing but blood and death. More than anything, he thinks, as he feels himself – his head – held aloft, that all he really wants is a drink of water. He decides that he will keep blinking to try and get someone's attention. It is to be his last thought on Earth.

Mohammed watches as Ali's head is separated from his body – watches his body topple over like a felled tree – watches his head held aloft, just like his father's. Ali blinks at him; Mohammed blinks back. Silence. There is nothing to say.

The camera captures everything – the high-quality digital footage is superb, as are the close-ups. But, best of all are the expressions on the two men's faces when they hear who will die. You can't fake that, you just can't – it is *so damn real.* That part will be downloaded and replayed all over the world millions upon millions of times in the coming week alone.

Rasmus knows that people are not watching the show just for the blood and gore and decapitation though. Oh no – there are plenty of images like that available on the internet already. They are watching it because it is they who are deciding who will live and die. They – the people – are all-powerful. They are God. They have the power of life and death over others – and it feels just great!

Mohammed is alive. But more than that, he now has what the majority of the kids watching him on their computer and smartphone and TV screens want and dream about, more than anything else in the world: fame.

Murderer of eleven, rapist of more, Mohammed is now famous all over the world. He is a celebrity, which is the best life can possibly get in the eyes of most young – and many not-so-young – people: a member of the new royalty, a new god in the pantheon, a hero. He is still alive too, of course, which is a definite plus.

He will soon be getting his prize money and a new life in The West. But that is just the start. Already, there are Facebook pages devoted to him; already, lucrative sponsorship deals have been offered; already, his future is secure. And all this when Ali's blood is still warm and sticky thick on the sand beside him, sipped at by buzzing flies, coagulating lifeless and sweet.

Mohammed has no idea why he is alive. He is, as he well knows, an unrepentant murdering rapist – and worse, he doesn't care. And yet the public – those infidels in The West – thought that his life was more worthy to be spared than that of the loving father, Ali (also a rapist and murderer), whose head now sits still on a spike in the middle of the courtyard, just like a pig's.

What is happening confirms to Mohammed every evil and violent misanthropic thought he has ever had about humanity. But if they want to give him money and a big house in Britain or America for taking part in all this, then why not? The West is full of slags to screw, for a start, and the money they said they'd be giving him – for ghost-written books and TV rights and even a movie deal – well, it is almost certainly more than a thousand generations of his oppressed ancestors earned, slaving away to survive in the dustbowl of the desert over the sad and savage centuries.

This is his destiny – he was meant to survive and Ali was meant to die. It is God's will. There must be a reason for his survival. Maybe Allah forgives him his wicked deeds? Or maybe even approves of what he has done? Surely Allah will not disapprove if he goes to The West and hurts the infidel there? Is not this written in the Koran by The Prophet, peace be upon him, that the infidel is worse than a dog and should be killed wherever a believer finds him? What happened was meant to happen: it is the will of Allah, who is the beginning and end of everything.

A buzz of blowflies is now swarming around the damp, blood-coloured sand of the execution yard, attracted by the irresistible sweetness of death and decay. Mohammed knows that it could have been his blood, drying from red to black in the sand. But it isn't, and Mohammed is not dead – no, he is more alive than he ever has been, feeling every fresh breath as if it were the last in his lungs. And he will live his life full well.

*

Alicia McVicar stands triumphant on the stage. This is her moment, and she knows it. She feels like she is being loved and enjoyed by the whole world – it is the best shag she has ever had and it feels just great! The crowd – her people, her lover, her congregation – is hanging on her every word, screaming and cheering and shrieking, as the sound system pumps out its hypnotic hymn of hardcore dance music. She knows that this is her big break, that nothing will be the same for her ever again, that this loss of anonymity is like losing her virginity, but better - much, *much* better. This moment will define her career. This moment is hers. This MOMENT!

'Bring it on!' Alicia yells as the spotlights and lasers dance in the night sky.

'Bring it on!' she screams at the audience who are always hungry for more.

'Bring it on!' she shouts, her voice swallowed by the massive sound of the crowd.

Alicia knows this is only the beginning - there is no telling where all this will lead, where all this will end. From now on, anything is possible. From now on, this is her new reality, and a new Alicia has been created.

Celebrity gossip websites have already raised her rating - from C-list to B-list. Almost A-list, when you think about it - and she knows she'll be up there soon.

'Bring it on!' she screams, and the people, as one, scream her name back to her, grateful at being in her special presence and utterly in thrall to her power.

Alicia McVicar is famous at last.

Bring. It. On.

*

I am Rasmus.

Today I begin this blog which, in the future, when everything is complete, and the consequences are clear, will be made public as a record of what happened and how, and of how far we have come. This I do for you, the people. For the future.

X-TV will change the world, because the world will be changed, and the world is TV.

And TV is everything. For as it has been said:

TV is a monster
TV is a moron
TV is us

I am Rasmus, and this is my reality.

*

The First Day

Rasmus walks into the world, a man born anew. He has changed, as everything must, and as everything will. Everything will change soon – the whole world – and Rasmus Karn knows that he will change it.

He does not sleep, for Rasmus Karn has work to do, now that he has the opportunity, now that he knows where he is going, now that he has been reborn as a prophet of a new age.

This is the first day. Day One of the future of Rasmus and the world, though the world does not know it yet.

Rasmus walks along the same old familiar streets, but the streets do not recognise him – not now, not after everything. Because Rasmus is different now. He is *other.*

Days pass. Weeks. Longer. Time's arrow arcs to its target. Then, Rasmus is ready – ready to begin. Ready to change the world.

Rasmus sits in his office. Thinking. Outside, London pulses with the usual hopes and desires. The people are making money and losing it, loving and hating, giving and taking – wired and connected: with laptops, notepads, tablets, mobiles, smart phones – all the high-tech skeletal ephemera of our digital age. Always on.

And the always-on generation is out there. Waiting. And always wanting more, and more, and more.

X-TV will be a success. More than that, it will change the way we watch TV forever. Rasmus knows this, because he knows what human nature is: he can really *see*. He possesses that rare quality called vision – the art of seeing things invisible.

Thursday is here. Thursday is always here, with Rasmus. Thursday is a man Rasmus trusts - as much as he can trust anyone.

He stands next to Rasmus in his office on the top floor of X-TV HQ. Rasmus sits next to him, framed by a huge window through which a view of London - all concrete and glass and history - pans out widescreen on the skyline. Below them, far below, the river flows past, timeless and grey as the sky.

The office, on the top floor of a fourteen-storey block, is spacious, modern and monochrome. This even applies to the artwork on the walls from which the savage and grotesque characters in original black-and-white satirical prints by Gillray gleefully sneer out at a flawed world, as they have for two hundred years. This is the only art-work on the walls of this office. Rasmus knows that these prints say everything that needs to be said about human nature, and also that London has not really changed at all in the two centuries since Gillray's pen spat them out.

Thursday looks at the city sweeping below and then at Rasmus. He sees a slim man, early thirties, wearing a white T-shirt, black jacket, black trousers: the clothes are like a uniform, and it is all he ever wears. Rasmus never wears colours, but he can't stop his glinting grey-green eyes, calm and clear, from shining bright colour out at the world. The face is pale, the complexion fair. Slicked-back light-brown hair of medium length is combed back from a large thoughtful forehead - it is there, just under the hairline on Rasmus's left hand side, that the clearness of the skin is broken by a blemish. The scar of an old wound, though more of a burn mark really, where something has taken away the skin in one sharp slice, though it would seem most of this is hidden behind the hairline. Apart from this, the face is unremarkable. Rasmus looks, as he himself would willingly admit, much like thousands of other men of his age who work in the media. But they are not at all like Rasmus, despite appearances, and he is not like them either.

Rasmus has a quality, an aura, a certain indescribable *something*, that Thursday has seen only very rarely in other men who are born leaders. People will follow someone like Rasmus just as they follow messiahs and monsters and madmen in Africa and all over the world. It is a certain *stillness* – a strength, a charm, a charisma, call it what you will – a calm confidence that exudes ambition and success and belief. You just know by looking at him that Rasmus is a winner – a leader to follow. Whatever *it* is, Rasmus has it. Thursday does not have it. Nor do you, probably. But the few – the very few – who do are special. They can make things happen, and make people make things happen. Indeed, they can change the world, and sometimes do, as history attests. Thursday has not told Rasmus that he can see this quality in him. This is not unusual, because Thursday rarely says anything at all: he is a doer, not a talker. Life has taught him that this is the best strategy for staying alive.

Thursday is an alleged war criminal from Africa. He is someone who could well be accused of committing 'war crimes' – something that Thursday calls surviving and being a proud warrior, like his ancestors. He looks like them too, unsurprisingly: tall, athletic and skin the deepest shade of black, dark as a shadow. Thursday is from a failed state in Africa. If they had let Thursday clean the place up, it would have been better – despite the rape, massacres and atrocities – or maybe because of them.

That is all in the past now – 'Praise the Lord!' – as Thursday always used to say when raping and killing cockroaches in his homeland. He took certain measures before leaving Africa to hide his identity – measures that mean he too was born anew when he came to Britain – measures which included burning his finger prints off by dipping his fingers in acid. But Thursday has known far worse pain.

Thursday says little, but he is not paid to talk – he is paid to enforce, to obey Rasmus's orders and make sure they are carried out, to be loyal and get things done. He is an enforcer, as well as a protector. He enjoys his work. Rasmus has never asked Thursday about his past.

It is Friday today. Thursday and Rasmus are watching the news on TV. This is the only programme Rasmus ever watches on what some used to call 'The Idiot Box' – though it is of course no longer the box-like cathode ray tube crate that televisions used to be, but a sleek, widescreen, sliver-thin digital oblong hung on the wall like a mirror. Like most of those in charge of producing TV, Rasmus rarely watches it – apart from the news and snippets of X-TV's own shows, and he doesn't even watch them all the way through. But he sees enough to know what he needs to know.

It would be better, of course, if the victims were Westerners – statistically, young, blonde, fit girls are always best to get the ratings, which is why the newspapers are forever full of them, particularly if they get murdered. But, for the moment, these brown-skinned foreigners will suffice. X-TV correctly predicts that no-one will be all that bothered about some raghead criminal getting his head chopped off in some faraway lawless land about which we know nothing and care even less – not bothered enough to take any legal or other action, or to try to close them down, that is.

The beheading is scheduled for that evening, at peak viewing time in the UK. A live event is taking place, complete with big outdoor screen, at a secret location in London – which is so secret that thousands of cheering people will be sure to be there after seeing the posts on social network sites and X-TV trending on Twitter.

*

The world is falling apart.

Everywhere, humanity is in crisis, economies bankrupt, and political leaders greedy, corrupt, dishonest and despised.

Everywhere, young people have no purpose and believe they have no future in a world whose values are imploding with their economies.

Everywhere, people of all ages lose themselves in drink and drugs, and the pornography of debt-riddled consumerism.

This I have seen, and this I, Rasmus, have come to change – to save humanity from itself. There is no other way.

Power is a TV screen. For this is where true power lies today, not with politicians or the corridors of state bureaucracy. The people need and want and deserve that power – and I shall give it to them.

I am Rasmus, and this is my reality.

*

Marked as URGENT
From: Hugo Seymour-Smiles
To: Minty Chisum; Oliver Allcock; Lucinda Lott-Owen; Sangeeta Sacranie-Patel
Cc: Benjamin Cohen-Lewis; Penelope Plunch
Subject: Urgent Meeting

Good morning, everyone.

Just a quick email to say I have scheduled a meeting of the senior management task force, (as discussed earlier), for tomorrow afternoon at 2pm in order to discuss our response to recent events, particularly with regard to the new online television competition and the BBC's ratings, going forward. It will take place in my office: room 2124.

It is essential that all selected senior managers attend this meeting.

I look forward to seeing you all there.

Regards,

Hugo

BBC Head of Vision

*

Hugo

A noise, and Hugo wakes up. Traffic. Noise. Daylight. Headache. Life.

He has been napping, snug in the comfy swivel chair in his BBC office, as he often does of an afternoon. Today, as per usual, he isn't feel refreshed at all. Instead, as he fumbles for his glasses, he feels worried, and weary, and perhaps even a little ill. Just tired. And old. And bleary-eyed. In other words, just Hugo.

Twelve-hour days as a senior BBC manager did that to you, even though much of that time involved sitting around looking busy in meetings, or preparing for meetings, or writing reports on meetings, or having meetings about meetings to discuss meetings.

Hugo is the BBC's Head of Vision, a new role, which used to be called Head of Television, and he has always supposed it means he should adopt the role of a visionary – a 'seer' – someone who can peer into the future, see the proverbial 'big picture', and advise departments accordingly. But, if Hugo is honest with himself, not even he is precisely sure what being Head of Vision means, or, indeed, should entail, though he has a list of duties on his contract, which he carries out efficiently, though with little obvious enthusiasm. But at least the job title sounds good – very post-modern and full of 'synergy' – and maybe that's the main thing?

There are plenty of jobs like this at the BBC, in the multi-purpose multi-storeyed, management layer-cake structure of The Corporation, and there are lots of managers like Hugo too who have worked there for most, if not all of their lives. 'Lifers', they're called, and Hugo is a lifer. Sometimes Hugo thinks that he has been a BBC lifer since birth, or even before.

Soon, they are all there – a small selection of BBC senior managers: Hugo himself, of course, organiser and chairman; then the flame-haired Minty Chisum, former Head of Drama but now Head of Entertainment, whose brief includes commissioning reality TV shows; Lucinda Lott-Owen, Head of Business Strategy; Sangeeta Sacranie-Patel, Head of Diversity; and Oliver Allcock, Controller of BBC1. There are other managers who could and should really be present as part of this important task force – but, this being the school Easter holidays, most are with their kids in Tuscany, the south of France or similarly posh and pricey sunny climes, so Hugo will just have to make do with the childless and the dedicated.

As he always does just before meetings start, Hugo takes off his spectacles and wipes the thick lenses with a small, greasy-looking square of silvery-grey material he has just extracted from a little plastic case for this specific purpose. Everybody who sees him do this finds this habit – this visual nose-blowing – intensely irritating, but Hugo has no idea of this, possibly because without his glasses on he is as blind as an old mole, so has never actually seen their annoyance. Nature has, as Nature will, made him look rather like the little velvet-jacketed creature too – a resemblance increased by Hugo's nervous habit of sniffing and twitching his nose, then scratching it, whenever he gets nervous.

Hugo closes his eyes and dreams of green. Because what Hugo really wants to do, and had wanted to do for years, is to move to the country and indulge his passion for organic farming and gardening. Sometimes, it is only thinking of all the fruit and veg that he will one day be growing in the peace and quiet of that bucolic bliss that keeps him sane: all those potatoes and onions and carrots and asparagus and...

'You still with us, Hugo?' says a voice which seems to force itself into every corner of the meeting room.

It is the voice of Oliver Allcock, Controller of BBC 1, who is the last to arrive. He is late, as usual.

'Right, um, Oliver, um, OK, let's...'

'Starting on time, as per fucking usual,' snaps Minty Chisum, Head of Entertainment.

'Sorry for being a tad time-compromised, Minty,' says Oliver. '*My bad,* as they say on the street. Just been doing a bit of blue sky thinking, video-conferencing with the guys at kiddies' TV. They've got some seriously incredible ideas for programmes up there. Just incredible!'

'Fucking shame we can't afford to make any of them then, isn't it?'

Minty won't let go; Oliver ignores her.

Everyone is used to Minty's swearing. She has always enjoyed swearing, but it has definitely increased with the seniority of her role, maybe because the higher up the greasy pole one climbs, the more badly one can behave – and get away with it, whatever *it* is. A few years ago, after a more sensitive BBC employee made an official complaint about her offensive language, Minty made sure she got herself diagnosed with a 'disorder' by the BBC doctor. From that day forward, with a newly-acquired diagnosis of Tourette's syndrome (or, more specifically, Coprolalia, the 'shit-mouthed' version of it), she has effectively enjoyed official permission to swear as much as she wants, and all with the special protection of the BBC's Diversity Department, which takes protection of, and equality for, those with disabilities, disorders and syndromes very seriously indeed. So BBC colleagues have to be sensitive to her needs – or else.

Lucinda Lott-Owen, Head of Business Strategy, keeps her head down – she doesn't want her eyes to meet Minty's, because they both know what Minty says is true. The budget at children's TV has been slashed, just like that for all drama – one reason Minty got the hell out before it was too late: it's just so much cheaper to buy kids' shows in from abroad.

'So,' says Hugo, 'let's, um, make a start, um, shall we? Right, well, um, we all know why we're here, so, um...'

'They hammered us incredibly hard in the ratings, X-TV,' says Oliver. 'This is not good.'

There are several shakes of several heads. They all know ratings for BBC1 should match or exceed those of ITV at peak time, and should be way above any other channel – digital or terrestrial. And now they have been beaten by a new, upstart channel, broadcasting reality TV online.

'Not good, no...' says Hugo.

'Course it's not fucking good, Hugo – it's a fucking cunting disgrace,' says Minty.

'Yes, um, Minty,' says Hugo, 'but...'

'Incredible show though, just incredible,' says Oliver. 'The look on those guys' faces when the winner was announced. Incredible jeopardy.'

Jeopardy: the gold standard and the aim of every reality TV show – the end of the quest and the thing that makes great telly work: the golden ratio that just has to be there for any show to shine, for anyone to get hooked – the only way to retain viewers, to increase ratings, to be a success. Without jeopardy, the audience just wouldn't care what happened to anyone – who lived and who died – which is why every single reality TV show tries to graft it on to any format if it's not there already, usually accompanied with ominous music, heavy pauses and silent stares.

'Just what the fuck is X-TV and who's behind it?' says Minty.

'Ah now, you see, that's what I've, um, been working on...'

Hugo knows that he may well be a plodding mediocrity – in fact, he is absolutely sure that he is a plodding mediocrity – but he's a plodding mediocrity who knows how to do research. Or, more accurately, he knows how to get others – junior staff, interns, underlings – to do research for him. He is, in short, a *professional.*

Hugo stands up, clicks a button on a projector and a logo

appears on a scruffy off-white screen hanging from the wall opposite him. Everyone turns to face it.

'X-TV,' says Hugo, 'is, um, the, um, enemy'.

The previous weekend, he has watched the original version of 'The Dambusters' on DVD, (when not watching 'Execution Night - Live', that is), so something of the manner of a World War II RAF Wing Commander has rubbed off on him. It's all he can do to stop his voice morphing into some Churchillian whisky-soaked Harrovian slur as he prepares his men (and women) for battle on the televisual beaches.

The large X-TV logo looms over the room from the screen. It is a stark, impressive and slightly scary logo. The X is like a swastika - in red, black, white and silver - and looks profound, eye-catching, important and impressive. Hugo's teenage children, Amy and Tilly, who thankfully spend much of their time away at boarding school, tell him that X-TV's logo makes the BBC's own little limp logo look "gay". He reminds them that not only is their use of the word "gay" inaccurate and extremely offensive to homosexuals - even if everybody says it at school when referring to anything that's rubbish - but that the BBC's logo was designed by a top design agency, cost millions of pounds, and has won several awards too.

'It appears that, um, X-TV is broadcast online from, um, an island...'

'An island?' says Oliver. 'What - like Canvey Island, or the Isle of Sheppey?'

Minty's eyes roll in bored amusement in their superior sockets.

'Um, yes, um, no, um,' says Hugo, 'looks quite a nice, um, island actually, from the, um, photos, in the, um, South Pacific - or at least, um, it would be nice if it, um, hadn't, um, been contaminated, um, with radioactive dust from an, um, H-bomb test in the, um, fifties...'

Hugo clicks up an aerial picture of the island on the projector screen, then a still of the mushroom cloud from 1953 and various photos of the devastation caused.

'Yes yes fucking yes,' interrupts Minty, 'but c'mon – these X-TV cunts must be based in London – every other fucking media type is.'

By *every other fucking media type* she means everyone in TV – or at least, anyone who mattered.

'Except those poor bastards up in Salford,' snorts Oliver, in a reference to BBC's new flagship 'Media City', and everyone else sniggers too: they just can't help it.

'Salford!' sniggers Oliver. 'Incredible!'

When the laughter dies down, Hugo continues.

'Right, um, yes, Minty, they are, um, based in London – West London. Hammersmith, actually.'

A hush descends on the room. Its occupants all turn their heads to look out of the window of BBC Television Centre in the vague direction of Hammersmith.

'So X-TV is right fucking here on our doorstep?' says Minty. 'And we didn't *know*?'

'Incredible! So who's behind X-TV, then?' asks Oliver.

Hugo is prepared. He sniffs three times, scratches his nose and clicks an image onto the screen.

'This, um, is our man,' he says.

It is a grainy shot, of the head and shoulders of a man, thirties, slim and with slicked-back hair. It seems to have been taken surreptitiously as the man is waiting to get into a car outside X-TV HQ.

A frown forms under the scary scarlet hairdo of Minty Chisum. She has seen this man before – she knows she has – and some memory struggles to bubble up to her consciousness, like a stubborn fart.

'Rasmus Karn,' says Hugo, 'that's, um, the fella's name.'

'Khan? But he doesn't look Asian.'

'No, um, Karn – K. A. R. N.' spells out Hugo. 'Not Khan as in, um, Genghis.'

'Rasmus?' says Oliver.

'Yes, um, Rasmus. Unusual name, isn't it? Could be, um, Norwegian or, um, Swedish...'

Hugo has spent many an evening in the company of gloomy Scandinavian crime series, so knows all the Nordic names.

'Norway,' says Oliver. 'Incredible place, incredibly expensive, incredible fjords.'

Hugo risks a joke:

'I, um, actually, um, think they drive, um, Volvos, not Fords, Oliver'

Nobody laughs, except Hugo.

'But Volvo's Swedish,' says Oliver. Hugo says nothing.

'Rasmus? As in Erasmus?' says Lucinda – it is the first time she has spoken at the meeting. She remembers something about Erasmus coming up when she was doing her A-Levels at Roedean.

'One, um, supposes so,' says Hugo.

'Looks like a fucking slimy little snake to me,' Minty hisses, spitting out the words with venom.

'Now now, Araminta,' says Oliver, 'we're all incredibly hard-working media professionals, both at the BBC and the independents.'

Minty tuts; Oliver smirks; Hugo nods. Sangeeta and Lucinda say nothing.

Hugo notices that nobody has spoken for a good few seconds – they all just seem to be staring at Rasmus Karn's face on the screen, transfixed. He clicks a button and up pops a PowerPoint slide displaying graphs that compare X-TV's ratings and those of all terrestrial and internet TV channels. Colourful graphs and tables of statistics always look impressive, which is why Hugo always uses them in meetings, especially when he doesn't really know what to do.

Hugo senses a sigh of disappointment in the room – the kind of bored painful silent wail of a groaning sigh that he remembers from double maths on Wednesday afternoon at school.

It is going to be a very long day.

*

Rasmus knows the BBC. Because Rasmus remembers.

Rasmus knows that there will be one person there who will recognise him, one person who knew him in the past, one person who will hardly believe his transformation into the man he is now. It may take her some time, but eventually she will recognise him, and she will scarcely believe her eyes.

But I, Rasmus, am prepared. Thus, I can anticipate what will happen before it happens: I can predict the course of events to come.

One thing is certain: X-TV is only in its very early stages, taking small steps into the world of television, breaking down barriers, building on what has gone before. There is more – much more – to come.

It is good to be able to copy reality, but to create reality is better. That is what X-TV will do and that is what it is doing.

I am Rasmus, and this is my reality.

*

Reality Bites

The grey-green, all-seeing eyes of Rasmus scan the room. As always, he looks calm and collected, still as stone. His smile – a small, secure, tight-lipped smile, beatific and beautiful – shines out into the room, a ray of light and a beacon, honest and good, into a dark and deceitful world.

The X-TV production team are having a meeting. Rasmus does not like meetings, so they take place only when absolutely necessary – a rare thing indeed at any media organisation. The rich dark smell of coffee makes everyone present feel warm, secure, alert and positive, which is just the intention.

A meeting is taking place now so that Rasmus can monitor progress of all X-TV reality shows currently in production. He rarely meets all the X-TV producers together, or even individually: he communicates to the team via email, usually, and sometimes by phone or video link.

Rasmus is, as ever, courteous and welcoming to everyone. He has that rare ability, when greeting someone, of making them feel they are the only person in the room and that he cares - really cares - about them. The way he looks into their eyes, the way he speaks so gently and clearly, the way he sees.

He knows that the producers he selected were bored with producing the same old reality shows for the same old tired and tiresome TV companies, and that the very generous salary package and the opportunity for real power being offered by X-TV were things they would never refuse.

Now they have the real opportunity to change the whole landscape of TV in Britain and beyond. When they meet Rasmus for the first time, they know he is a man they can do business with. And he is.

All the X-TV producers are there at the meeting. None of them has met Rasmus more than once or twice before. For many of them it is more than a little disconcerting to be sitting there in the same room as the genius who created X-TV - the man who started it all, and a true visionary. They remind themselves that he is their boss, the creative spark and guiding light of X-TV: Rasmus created this world, from nothing, or so it seems. It is he who had the vision from the start; it is he who has the vision now; and it is he who is the creative force who will change the world of TV forever.

All the senior producers are there: Debbie Owen-Tudor - a thirty-year-old hippy-chick producer of cutting edge documentaries, usually about sex or disabled/disfigured people or both, and who has previously worked for Channel 4; Mercy Matumbe - smart, twenty-five year old daughter of a Nigerian diplomat, whose contacts got her a job at the BBC, where she spent the previous four years; Anita Bergman - half-Asian, half-Swedish, and another one who jumped from the BBC ship; Gary Wu - half Hong Kong Chinese, half Cockney, who worked on three series of Big Brother as a senior producer, while still under thirty; Ravi Govinda – the main finance man, still only twenty-three, who loves a spreadsheet more than any woman he's ever known, possibly including his mum; and Sebastian von Saxonburg – young, rich and aristocratic, and for the last two years, since leaving Oxford under a cloud of cocaine, highly successful producer of cutting-edge shows about celebrity, drugs, fame, drugs, sex – and not forgetting drugs.

Ravi and Anita are the main producers responsible for 'Execution Night – Live'. Consequently, they look proud and perhaps a little smug at the meeting. Ravi has just flown back from Afghanistan. He seems tired but slightly startled at the same time. Anita, who produced the event hosted by Alicia in East London, looks even more startled – but then the more vibrant parts of East London can be shockingly violent these days.

Rasmus sits in his usual large black chair. Thursday is standing by his side – guard, protector, servant, admirer, friend. On assorted seats around a large table sit the X-TV producers, most of them dressed in smart-casual suits in the TV producer tradition.

Rasmus is reading out the ratings for 'Execution Night – Live' and those for the other TV channels that week. When he reads out the ratings of BBC1 – which are well below those of X-TV – there are nasty laughs. It feels good to be the much-talked-about upstart, the new kid on the block, the mover and shaker – and it feels even better to be the underdog when you're biting back.

'Welcome, everyone, and thank you for coming today' says Rasmus.

He smiles to the room and the room smiles back.

'Many congratulations are due to everyone here for making 'Execution Night – Live', only the second X-TV production, such a great success – but especially to Ravi and Anita.'

Everyone glows with pride: they have all seen the initial ratings. Ravi and Anita bask in their colleagues' praise, but with decorum and modesty, happy and proud at their success. It is a long time since either of them has felt so good – so fulfilled, respected and appreciated – in a TV production meeting.

'The rats, please,' says Rasmus to Thursday, his voice quiet and smooth as a pebble.

Thursday selects a PowerPoint slide from the laptop and a bar chart appears on a large TV screen on the far wall of the room.

For a few seconds, everyone – (including Rasmus, who allows himself a brief moment of emotion) – gloats inwardly at the success shown by the snazzy graphics: X-TV's ratings are represented as a tall and erect bright red bar – it is taller than the others. Next along, the BBC's blue bar which is noticeably shorter – the space between them represents the 4.7% higher audience share that X-TV got for 'Execution Night – Live' Friday night.

Way behind, the other terrestrial and digital TV channels, many too tiny to be more than a barely visible line on the bar chart. No other online TV channel has registered enough viewers even to have a bar on the chart to themselves; instead, a small bar labelled 'Other Online Channels' represents their combined minuscule viewing share.

Thursday presses a key which starts a film show detailing how the ratings changed in the week leading up to 'Execution Night – Live'. Everyone watches as BBC1 and ITV start strong, way ahead of any other channel, but then the little red square labelled X-TV grows and grows, heading ever-upwards like a time-lapse skyscraper being built, as both BBC1 and ITV's bar charts steadily shrink and diminish day by day until – at last – that magical moment arrives: X-TV overtakes the BBC on the final day. For the first time in broadcasting history, an online TV channel has beaten all terrestrial channels in the ratings.

Suddenly, and spontaneously, there is a small round of applause in the room, some clenched fists punching the air accompanied with a 'Yes!' or two, and various assorted laughs and cheers. Debbie and Anita hug; Gary pats Ravi on the shoulder; Sebastian is bored with envy.

For a moment, everybody just gawps at the screen, transfixed, hypnotised and struck dumb – which is ironic really, as this is what viewers tend to do too when they watch reality TV: it is the primaeval yawn of Early Hunter-Gatherer Man basking in the glow of his campfire after a kill and staring transfixed into the flames.

'So, all I can say is well done, people – but, as we all know, we can always do better. We shall make X-TV the biggest television company in the UK,' says Rasmus, as if stating the inevitable fact of an eternal truth. 'We can do all this, and more. We can be the best and we shall be the best. That is what I want everyone here to aim for. Always. Because there will always be those out there trying to bring us down.'

'Sex and violence,' says Sebastian sleepily, with his eyes closed and head leaning back, 'are the new gods...'

'Smashing down barriers, breaking the mould and making cutting-edge programmes,' says Rasmus, 'is exactly what we are here to do.'

Everyone nods in agreement. They have all been working on shows that will do just that.

Rasmus smiles: he knows his team will not let him down.

*

The individual is all.

If history teaches us anything, it is that all achievements of human civilisation, all inventions and improvements, all creativity and genius, have always sprung from the minds of the few lone individuals who are not prepared simply to follow everyone else – to follow the crowd, the herd, the mob – but instead have their own vision of just what can be achieved. There is always just one man at the start and heart of everything.

Without Rasmus, there would be no X-TV and the media

landscape would not be undergoing radical alteration. It all starts with one idea in one mind.

However, one man cannot achieve everything by himself, despite creating a vision: others are needed to put any plan into action. Leaders need followers to spread the word and make a vision a reality. Inventors need other people to improve on, develop and manufacture their inventions, and to market and sell them. Writers, painters and musicians have similarly unique and individual talents, but they need others to make their vision a reality communicated to the whole world. It cannot be done alone, though the vision – the true start and inspiration, the spark that lights the fuse – must be that of one individual. Vision has never been a team game, and visionaries are not team players.

Thus I have gathered around me my people – my team – my disciples. These are the tools I need to create what I have to create.

These individuals will make my vision a reality, and for doing so will be well-rewarded: after all, TV is not a charity, and the people who work in it expect – and demand – to be well paid and valued. I pay this to reward total loyalty to my vision – to X-TV and its future.

And in that future we shall create a whole new world.

I am Rasmus, and this is my reality.

*

The Presenters

Presenter 1:

'Yeah cool right yeah wicked!' says the very young and very heavily moisturised shiny-black baby-face on the screen.

'This,' says Rasmus, 'is Danny Mambo.'

'*The* Danny Mambo, yah?' says Mercy.

'Who's got that black music show on the BBC?' says Ravi.

'The same,' says Rasmus, 'Danny works for us now'.

'Well cool for getting the kids who's well into black music, innit?' says Gary Wu.

Danny Mambo has a zig-zag design carved into his hair, and gold and diamond-studded teeth which glint when he grins. His hands make ghetto-gangsta-rap shapes as he speaks, emphasising the words like a rapper, though he resembles, more than anything else, a very young Louis Armstrong.

Presenter 2:

'I is da man, d'ya get me, I is *da man!*' says the next presenter to appear on the screen: a short young man with a very pale brown face, a big afro haircut, a huge pair of sunglasses and enough jangling gold on his wrists and fingers and around his neck to pay off the debts of a small third world country – and perhaps even one or two European ones.

'Nazish Naidu, experienced TV and radio presenter,' says Rasmus.

'Oh I can't stand that wanker,' says Ravi, almost before he realises he's saying it.

'Why's that then?' asks Gary. 'I worked wiv the geezer before – he's well cool.'

'Look, he's British Asian, this Nazish guy, right?' says Ravi.

Gary shrugs: 'Yeah, s'pose. So?'

'So, in that case, if he's British Asian, why's he trying to sound like a black man from the ghetto. No offence, Mercy.'

'None taken, yah, sweetie,' Mercy grins.

'Nazish is a very popular presenter, amongst a young C1 and C2 demographic, yeah, Ravi,' interrupts Debbie.

'Oh I know I know,' says Ravi, 'and it's cool, if the kids like him...'

'Yeah mate, too right they do,' says Gary, and a quick scan of nods and raised eyebrows around the table shows all others present are in agreement. Rasmus enjoys the now unanimous approval of his choice of presenter.

Nazish raps on, in his weird pretend half-African-American half-Cockney gangsta accent, which belies his Wolverhampton roots. He has worked hard to eradicate all trace of his Brummie accent, mainly by listening to rap records, though the occasional 'yow' (instead of 'you'), and over-pronounced 'g' (in an occasional 'ing') still give him away. But nobody really notices, because what he is saying seems to consist of the same dozen or so words, endlessly repeated, in various orders:

'I is da man – *da man* – d'ya get me, bruv – rispek – jah see now, innit – rispek!'

Presenter 3:

Thursday fast-forwards to the next presenter, to Ravi's obvious relief.

Onto the screen there now bursts a young woman screaming like a lunatic and shaking around her wild head of hair – and her breasts – like a banshee possessed. Eventually, the girl stops shrieking and actually manages to say something intelligible.

'Hiya! Bring it on! Yay!!!' she squeals. They all know her – it's Alicia McVicar, and this is her audition tape.

Everyone can see that the young woman on screen comes across as a loud, irritating, attention-seeking, airheaded, noisy bimbo with no real talent or ability whatsoever. In other words, she is absolutely *perfect* for TV.

Presenter 4:

Thursday grins as he fast-forwards to the next presenter.

The screen now shows the face of a very young and angelic-looking blond boy, with large blue puppy-dog eyes of the sort that you just want to dive into forever. He has perfect, lilly-white, blemishless skin which almost glows like the finest porcelein.

'Wow, I mean, this is like reet good,' says the shy boy.

His accent is broad Sheffield, and his voice soft and gentle, not shouty like the other presenters. He seems as naive and innocent as a young child – or perhaps a little lamb.

'Coz like... it's all I ever wanted to do, like, yeah... present shows on't telly and that...for me mum, like...'

'Now *he* is a sweet boy,' says Sebastian, 'so...Caravaggio...'

'He is, yeah!' says Debbie – feeling both protectively maternal and sexually attracted to the boy at the same time which is, she knows, the reason why he has been selected as a presenter in the first place. He'll appeal both to the young dippy girls and the sad ageing maternal women at the same time – not forgetting the gay demographic, of course.

'A definite sweetie, yah,' says Mercy, grinning widely to reveal her huge tombstone teeth.

'Does this angel have a name?' says Sebastian, melancholy with lust.

'Don't you recognise him?' says Rasmus. Everyone looks again. Thinks. Looks again. And then it dawns.

'Course, I seen that geezer before' says Gary. 'Weren't he a contestant on one of them talent shows?'

'Oh yes, I remember him too, last year – no, two...maybe three... years ago, yah?' says Mercy.

'Not Pop Idol, yeah,' decides Debbie, who's a big fan.

'Totally not on Britain's Got Talent either,' says Anita, who worked on it.

'What the bleedin' 'ell's he called again?' says Gary.

'Am I missing something?' says Ravi, baffled. He has always hated talent contests, ever since his mum made him enter one when he was a kid, dressed as a Channel 4 newsreader.

'Colin...something?' suggests Debs, but she knows it's wrong.

'His name's Calvin,' says Rasmus. 'Ring any bells, people?'

Thursday pauses the footage. Everyone in the room looks at the boy's blushing face frozen on the screen in mid-mumble.

'Pop Starzzz,' says Rasmus. 'Series five.'

Recognition of Calvin's name – and face – gradually takes shape in the minds of the assembled producers, the muddy waters of transient memory gradually clearing to reveal a televisual jewel sparkling in the silt.

'Calvin – now I remember him,' says Debbie, with a sigh of relief. She is amazed at how quickly she forgot all about him after he disappeared from Pop Starzzz.

'Yah, me too,' lies Mercy.

'Totally,' says Anita.

There are nods around the room.

'This,' says Rasmus, 'is Calvin Snow.'

'Pure as the driven, I don't think,' says Sebastian.

'Who was favourite to win Pop Starzzz series five, yeah?' remembers Debbie

'Until it was revealed that he was, in fact, sixteen and not the eighteen year old man he claimed to be, the naughty boy,' says Sebastian, remembering.

'Absolutely, so,' said Ravi, 'now he must be...'

'Eighteen last week,' says Rasmus.

'And my, hasn't he grown?' sighs Sebastian.

'So, people,' says Rasmus, turning off the audition film, 'those four are X-TV's initial presenters'.

Every one of these four presenters will soon be mega-famous – not famous as a loser on a reality TV show, or as a presenter of documentaries no-one watches, or as a DJ on late-night rap music radio shows with tiny audiences. But *really* famous: so famous that everyone in Britain, and perhaps much of the world, will know their faces and names. It is what they all want. It is all they always want. Rasmus sees that wide-eyed yearning desire in their eyes as he has seen it in the eyes of so many others, the tell-tale sign of a dreadful and relentless disease: the desperate desire to be famous.

And fame is what these four will get – and then some too.

*

Nobody lives forever. This is fact, as far as we know.

And yet, it is possible to live more than one life, to be born anew as a different and changed being. Some claim this happens in religion or spiritual awakening, that sense that all that has gone before is merely a preparation for rebirth, a time when the real being – the true man – is being incubated, awaiting the day when he can be reborn.

In the past I would not have believed such a phenomenon could exist. Now – now that I know what it is like to be reborn – I do.

I changed one night – the night I left my home with no intention of ever returning – the night I was going to end my life. Before it happened, before I could throw myself to my death, a lightning bolt struck from the heavens, its electricity jolting me back into a living being.

But the past is the past. Never look back. Live in the now. And create the future.

I am Rasmus, and this is my reality.

*

Email - Marked URGENT!
From: Hugo Seymour-Smiles
To: Minty Chisum; Oliver Allcock; Lucinda Lott-Owen; Sangeeta Sacranie-Patel
Cc: Ben Cohen-Lewis; Penelope Plunch
Subject: New Meeting Re. Ratings Task Force

Good morning, everyone.

Thank you all for attending the meeting last week. Most productive and interesting, with some fantastic blue sky thinking.

I hope everyone has had a chance to digest all the information by now, so I'm scheduling a meeting for 17th April, again at 2pm. Please let me know if you cannot attend.

I look forward to seeing you all there and hearing more of your contributions, going forward.

Regards,

Hugo

BBC Head of Vision

*

Minty

Minty knows Rasmus. She knows she knows him. She just *knows*. Their paths have crossed before – but where? And, perhaps more importantly, how? What is the context; what is the connection? Minty Chisum just can't remember and hates herself for it.

She has met so many people over her years at the BBC, both as Head of Drama, and now as Head of Entertainment, and she prides herself on her memory for names and faces. She knows from experience that actors tend to be such vain and preening narcissists that they always need to be made to feel as special and talented as they consider themselves to be. Therefore, remembering such people's names was absolutely essential to boost their fragile egos.

Minty Chisum knows for certain that she has definitely not knowingly met anyone called Rasmus before, and the only 'Khans' she has met have been minority ethnic. Maybe he – whoever *he* is – changed his name? But that face – she knows she knows that face, though it looks different somehow... But she doesn't remember ever meeting someone with a scar like that on his forehead – she would have remembered that, surely? But she can't remember, not this time. Why can't she remember him – whoever *he* is?

But whoever Rasmus Karn is, he is the enemy and has to be

destroyed – because his destruction will mean the destruction of his company, X-TV, something that Minty sees as absolutely necessary if the BBC is to retain its unique position in British broadcasting and society. She knows this has to be done, and that doing this will also make her a prime candidate for the job of Director General, when the present incumbent, Benjamin Cohen-Lewis makes way.

Minty is sitting on a retro Habitat sofa which rests on the wood-panelled floor of the living room of her townhouse in Hammersmith, looking through the documents prepared by Hugo and his colleagues. There is a list of names on one document: the X-TV staff. There are photos too. Hugo didn't want to distribute these to those present at the meeting – (it would have seemed inappropriate somehow, as if he had stooped to the depths of the paparazzi) – but he certainly wasn't going to argue with Minty when she asked for copies. The first thing Hugo did when he got back to his office was to email her the JPEGs.

'I finish,' says a small and weary-sounding Slavic voice. It is Oksana, the cleaner, who comes twice a week.

'I *have* fucking finished,' shouts Minty without raising her eyes from the photos.

'I heff fucking finish,' says Oksana.

She stands in the arched doorway, her face blank and uninterested. In her pocket sits the little envelope of money from the kitchen, which is all that matters to her.

'I sorry, I go,' she says. 'I come Thursday.'

'No – Tuesday – fucking *Tuesday!* – not fucking Thursday!'

Oksana always gets those days mixed up.

'Is what I say – Tuesday,' she says. But what she is thinking is something entirely different, namely: *'Go to the hell, ugly red-haired pig woman!'*

'Fucking potato-munching Polish slag,' mutters Minty to herself as the front door slams, though Oksana would tell her, if she ever bothered asking, that she actually hails from Ukraine. Minty has got through so many cleaners over the years, and they all sound Polish, so that's what they are as far as she is concerned, whether they like it or not.

Minty looks at the photos. It seems Hugo is not completely useless after all – because he has somehow managed to get shots of all those leaving or entering the X-TV offices. It disconcerts Minty to think that these were taken less than a mile from where she is now sitting. Minty prides herself on knowing everything about everything and everyone in the TV business, about being on top of everything, always. But she is well and truly out of the loop this time, and she hates it more than she can say.

A black face stares out at her from a photo – it is the only one that seems to be looking directly at the camera. But that directly-looking-at-the-lens is only in one shot; in the others, the man is looking elsewhere – that must mean he's being very vigilant, she thinks. A bodyguard perhaps?

Together with the photos, Hugo sent Minty a list of X-TV's senior producers and details of their TV careers thus far. Some of the names register in Minty's mental database of media names – she remembers most of these people, however vaguely. All younger than her of course – much younger. Too young.

Minty Chisum is fifty-six years old, though her bright red hair and the outfits she wears make her seem ten years younger. She makes TV for young people, so considers it essential that she is on their wavelength – that she knows what they like and dislike, and how they think. There is, happily, a perpetual population of noisy sexy youth rising up to meet her, and then when those kids grow up and pass by on their way to world-weary adulthood, there is another fresh crop coming up behind, sap rising in readiness to be tapped by a TV producer like Minty.

But fifty-six is not young – it's not even on the edge of young. Fifty-six is old – nearly retirement age. That is what Minty realised a few years ago, and that is why she had to act before it was too late.

For far too long, Minty knows, she stayed in the same

job as Head of Drama, and spent years before that in various sections of that department and others, using her sharp elbows to nudge – and slice and slash – her way to the top. Then, after years in the job, Minty Chisum had her epiphany. And lo, verily, she saw what others had seen before her and which others are yet to see – that TV drama, on the BBC or any other channel, is more or less finished, and will certainly never again matter in the same way it once did. She saw, in other words, that there is far more drama in reality TV – whether "constructed reality" shows or any other kind – than there is in any actual drama on the BBC or anywhere else; that it is reality TV which created the kind of cutting edge drama of the type that hooks a nation and tells the story of who we are.

She may well have come to it late, and for far too long she sneered and scoffed at the awfulness of reality TV, like so many others. But as a convert to a form of television she previously despised, Minty Chisum is now its strongest advocate. Reality TV is the future and will be forever: its fiction – its storytelling – its sheer enjoyability – is better and more profound than anything in TV drama departments these days.

Of course, the viewers who love, adore and worship reality TV and its stars don't care which company actually makes it. All they care about is whether or not it gives them the opportunity to be entertained, and especially to laugh at and feel superior to the losers on the screen and thereby to feel better themselves. It is pure meritocracy – power to the people – a truly inclusive and egalitarian phenomenon.

And this, Minty knows, is what reality TV's critics simply cannot abide: having their precious television taken over by the people themselves and not remaining in the sole control of that aloof and haughty high-minded elite that it used to serve so well. Minty knows that the BBC has to give the people what they want. If that doesn't happen, then there is no future for The Corporation.

One big problem, as far as she can see it, is that, unlike X-TV, the BBC does not own an island in the South Pacific from which to broadcast its reality TV shows - shows that would break the law if broadcast from Britain or any other Western country.

And the BBC is not just another TV station or company. The BBC is special and always has been special. It is seen as special by the public - almost like royalty - and they thus expect higher standards from it than from other channels or from an online broadcaster. But then, the same public who expect certain standards of taste and decency and quality from the BBC also spend hour upon hour watching downmarket salacious reality TV shows on other channels, and constantly demand more violence, sex, perversion and humiliation from these shows too. It is a difficult circle for the BBC to square. All attempts to do so thus far have come out as often embarrassingly shapeless squiggles.

So what can the British Broadcasting Corporation, the world's first national broadcaster - whose duty has been to inform, educate and entertain the nation since its inception in 1922 - do in such a situation?

The answer is as simple as it is predictable: panic.

Home with Hugo

'What on earth, um, are we going to, um, do?' says Hugo to his wife, Felicity, who is loading the dishwasher in their expensively granite-worktopped Twickenham kitchen.

Felicity hates the way Hugo says "*we*" when talking about himself. As if Felicity had anything to "*do*" with anything. Typical man! Why on earth should she care if a new TV company was running rings round the BBC? She has her house, her children and her own busy schedule - of shopping and lunches and charity events: she is a very busy full-time mum! All she knows is that her husband earns a good salary. Her interest ends there.

'I mean, um, we have to, um, compete with the online TV channels but, um, how, um, can we?'

Felicity closes the dishwasher door and presses a button. It whooshes and slurps into action.

'When our hands are tied by, um, rules and regulations?'

Felicity walks out of the room without a word, but Hugo continues talking:

'I mean, it would be, um, super if we could, um, plumb the depths, as the online channels do, um, but...'

The dishwasher hums and sloshes away. Hugo is talking to it – a habit he has acquired at some time in the last few years (he can't remember when exactly) when his wife and children began to completely ignore him. These days, they have stopped even pretending to pay attention when he speaks.

So, instead, Hugo pretends that he is talking to some imaginary person in the room – some understanding wife, perhaps, who loves him and is just a little interested in his life and career. This means that his conversation at home is often aimed at the focal point in the room – the dishwasher, cooker or fridge. He talks to these more than he does to anyone or anything else, (including the tree in the garden). And they talk back to him too, in their way. Just listen to the dishwasher purring its approval right now! They are, perhaps, his best friends.

It is they – and they alone – who have been made party to the knowledge that what Hugo really wants is to leave the BBC and become an organic gardener-cum-farmer in the countryside. Neither his wife or children, nor anyone at the BBC, has any idea of that ambition or just how keenly Hugo is counting the days until retirement.

Minty in Charge

Minty is not happy – not happy at all.

'We can't fucking use these fuckwit ideas,' she says, 'they're so...'

'Incredibly derivative?' suggests Oliver.

'Insipid,' says Minty.

Insipid. Hugo rolls the word around in his mind, sucking on the salty pebble of bitter criticism.

'Insipid...wet...just *so fucking BORING...*' mock-yawns Minty.

She is interrupted by the sound of muffled dance music. Lucinda takes her mobile out of her bag.

'Sorry,' she says, clicking it off, not needing to see the name of the caller: it's her bank again. Minty looks daggers at her.

'So, um,' says Hugo, 'does anybody else, um, have an opinion?'

Always ask for people's opinions, thinks Hugo. It shows you're interested, for one thing, and it also conveniently covers up the fact that you have no ideas of your own.

Oliver can't help himself.

'Gotta agree with Araminta here,' Oliver says, 'they're all incredibly bloody boring ideas.'

They have all just been going through various programme ideas from the BBC creative team – Hugo asked a selection of the BBC's "talent" to throw some ideas around after the last meeting. It is a list of ideas which, even Hugo has to admit, are piss-poor copies of programmes which have already been made, by the BBC and others. Not one of the ideas is original; not one of them will make good – or even passable – telly. They are all, quite simply:

'Bollocks!' says Minty. 'Complete and utter fucking bollocks.'

Hugo winces and twitches the itch on his nose.

'So,' says Hugo, 'have we, um, by which I mean you, and, um, me, got any other ideas for, um, programmes?'

Minty stares at Hugo, her face darkening to a shade of bloody red to match her hair. She is a creative professional, for goodness sake – and now some manager who has obviously never created anything but chaos and confusion and crap is questioning her creativity, after all those years she spent at BBC drama!

Oliver Allcock, Controller of BBC1, breaks the silence.

'Everyone here's got incredibly good ideas, Hugo, which is why we are all where we are, at the BBC,' he says.

'Yes, um, but...'

'The creative team here is the best in broadcasting – incredible people, with incredible talent, and incredible ways of doing incredible things – incredibly!'

Everyone nods, but everyone also senses the horrible hollow irony that the BBC's creative team – its best-paid producers, the cream of British creative talent, highly educated and dripping with confidence in their abilities – can't come up with a single worthwhile and workable, half-decent creative idea.

'Perhaps Hugo has some *other fucking ideas* for programmes'?' says Minty.

'Um, well, that's not what I...'

'Oh but surely Hugo, as BBC Head of Vision you must be able to fucking well *see* something!' says Minty.

Hugo's humiliation is as complete as it is predictable. In some peculiar way that he simply can't fathom, he has somehow managed to unite the bitter enemies Minty and Oliver against him. He has no idea why, but his BBC meetings always seem to turn out like this. It's always happening these days. How? Why?

'Well, um, no, actually...um...not really, no,' he says.

'What a fucking surprise,' says Minty.

'Incredible,' says Oliver.

Hugo sniffs like a schoolboy and scratches his nose.

Sangeeta and Lucinda nod at each other. They can both see that it would be dangerous to support Hugo, an 'old man' who will soon be retiring anyway. For if they did, then they could be dragged down with him when he is finally sunk, and neither of them would want that – not now, not when they've worked so hard to get promoted to a level so far above their natural level of competence.

'I didn't mean to suggest that, um, nobody here had any good ideas,' says Hugo, knowing full well that this is precisely what he had been suggesting.

Minty ignores him. She knows that he is useless – like most, if not all, men – though she can remember a time, long ago, when Hugo was confident, efficient and good at his job, before marriage, before everything, broke him. So she knows a woman will have to take the initiative here, take control of the situation and sort the mess out.

'Oh, *fuck* this for a game of cocks and cunts,' says Minty, banging the table with her fist. 'We need Peter fucking Baztanza – it's the only fucking way.'

A hush descends upon the room.

Sir Peter Baztanza: a man who has done more than anyone else in the previous twenty-five years to dumb down British TV. Practically every single hit reality TV show from the previous three decades has Baztanza's grubby fingerprints all over it, from those that put strangers into a situation together – a house, an island, or a jungle, perhaps – to those that are, frankly, perverted and voyeuristic, such as intimate and gory cosmetic surgery programmes – to the enormous amount of property and food porn on TV – and many a TV talent show too. In fact, the favourite programmes of much of the British population, including Sangeeta and Lucinda, began life buzzing in the ratings-hungry always-on brain of Peter Baztanza. Needless to say, he is also a very wealthy man indeed, having sold his programme ideas to almost every country in the world. At any time of day or night, programme formats that he created are being broadcast somewhere, earning him revenue 24/7.

'But we can't,' says Hugo, 'not, um, Peter, um, Baztanza.'

'Yes, we fucking can,' says Minty.

'It'd be incredibly difficult to get him – it'd have to be an incredibly attractive deal.'

'I fucking well know that, *Ollie*,' snarks Minty, knowing full well how Oliver hates to be called that, 'but it's the only way to get all those brain dead bastard viewers out there watching the BBC again.'

'Incredibly important,' says Oliver. 'Hit the nail on the head incredibly hard there, Araminta.'

Eyeballs meet eyeballs in silence. They know the BBC can always find the cash when it really wants to, despite its constant moaning and whingeing about funding. But it'd perhaps be harder to justify to the public the paying of an enormous sum of cash to an already extremely rich TV producer from outside The Corporation to create programmes that the BBC, with its thousands of employees and highly educated workforce – *and* its four billion pounds per annum budget, should be well able to create all by itself anyway.

'But, um,' says Hugo, 'can we, um, really do this?'

Everyone looks at Minty; no-one is sure, but no-one has any better ideas for improving the BBC's ratings.

Minty knows they don't really have a choice.

*

Rebirth. Renewal. Change.

Parthenogenesis. The creation of life from within. The shedding of old skin to be born again as a new being. A kiss to wake the dreaming bones.

This is what happened to me, and this is a process every individual and organisation needs to go through to grow – to become another.

It is survival of the fittest: though it is not the strongest or the most intelligent who survives, but the one most adaptable to change.

A new, fresh, young company can move faster than any old, slow-moving beast, set in its ways and resistant to change – and that, plus creativity, focus and knowledge, is what will make X-TV successful and strong.

I am Rasmus, and this is my reality.

*

New Season

'Suicide Bomber Week – Live' was turning out to be an absolute blast.

It made sense that, as X-TV's production staff had worked

hard to make contacts in Afghanistan – (something that had led to the kidnap and violent deaths of two local interpreters) – then they might as well use the location again for the next show. Arrangements had been made with all authorities, deals struck with a few warlords, and permissions granted for programmes to be made.

The public are ready for what they are about to see. They've voted in their millions during 'Execution Night – Live', and the clip has been downloaded online many millions of times too. In fact, more people voted for their chosen outcome in this TV show than cast their ballot in the last General Election, especially those in the all-important under 25 demographic.

Local young people volunteer in their droves to take part in Suicide Bomber Week. Ten are selected, of whom five are guaranteed to die horribly and violently, blown to pieces by a vest packed with explosives. The survivors will get cash and a new life in The West – and at those 50/50 odds, the Kabul production office is inundated with thousands of applications.

There are complaints, of course, that a TV company is planning to broadcast such an offensive and appalling programme. The usual hand-wringers – from across the political and religious spectrum – object, as they always object, to anything groundbreaking and new. But there is nothing they can do.

'Suicide Bomber Week' is to be the last X-TV foray into Afghanistan, however – for the moment at least. Market research has indicated that, although people do enjoy watching live broadcasts of stomach-churning horrors occurring in such parts of the world, and particularly relish having a vote as to who will live or die, the fact remains that the victims are poor, brown-skinned and foreign, even if they do speak some pidgin English, so their deaths are worth less, televisually speaking, than those of white Westerners. For most viewers, it is almost like watching another species being killed, because the lives of the victims are so far removed from their own. After all, what does some teenager in Newcastle or Norwich or Newport really have in common with someone of the same age in Afghanistan? Everyone in the media knows this. The best people to suffer on reality TV shows – the ones which really draw in the viewers – are of the shade commonly referred to as 'white'. If they are young, blonde and female and well-endowed, all the better. And if the people on TV are already famous – or at least known as 'celebrities', perhaps from other reality shows – then better still: viewer recognition is always a plus.

Rasmus knows this, so new shows with Western contestants have been in production for months: it is a policy of X-TV to be several steps ahead of the competition and to always have several shows at various stages of production at any one time. They will never buy in anyone else's programmes, so need to ensure that if one of their shows is an unexpected flop, there's always another one ready to be broadcast. Not that Rasmus expects any of X-TV's shows to fail: he knows human nature well enough to be able to sense exactly what the people want. But he also knows it is best to expect the unexpected – and to have a strategy ready and waiting in case of unexpected failure.

'So, people,' says Rasmus, 'any questions?'

There aren't any. Everyone knows in detail what their roles are; everyone knows exactly what they are hoping to achieve; everyone is focused and ready. X-TV is a well-oiled machine with precision engineering.

'In that case, let's have updates about all shows in production. Debbie?'

Eyes turn to Debbie Owen-Tudor. Rasmus leans back and listens.

'Well, Granny Gang Bang is all set to go live tomorrow. Ravi and I will supervise that.'

'Nice one,' says Gary.

'Presenters will be Alicia,' says Debbie, 'and Calvin Snow.'

'Oh please keep that poor boy safe from that...thing!' says Sebastian, eyes dark with disgust.

'Nazish and Danny will be presenting Suicide Bomber

Week all this week, and that's followed directly by Granny Gang Bang.'

'For a crossover audience, yah?' says Mercy.

'Awesome! That is just so totally cool,' says Anita.

'Granny Gang Bang' is X-TV's first sex show, though there are many more to come. In it, various ladies aged sixty and over will be ready and available for sexual activity, as will some young porn models. Various attractive young men have been selected to take part, from many thousands of applicants, and the public vote will decide which of them will have sex with the grannies and which will have sex with the young women. There are forty young men there and it will be a fifty/fifty split: half will get the fit younger models, and the other half will have sex with women old enough to be their grandmothers. Crude, but effective, and guaranteed to disgust the BBC and many others even more – and to get massive ratings too.

Also planned for X-TV's New Season are the following:

'Execution Specials', with the public being able not only empowered to choose *which* criminal in some god-forsaken Third World country will die, but precisely how. Imaginative possibilities include being boiled or burnt alive, bitten by snakes, bled to death, drowned by a rising tide, eaten by eels, dissolved in acid, ripped apart by horses, and many *many* more eleborate endings to insignificant human existence.

Then there is 'The Dare' – in which members of the public are dared to do such things as eat live foetuses or faeces, and much else besides. There have been similar shows on TV before, but the dares have been tame: X-TV's 'The Dare' would take everything to a whole new level.

And, bearing in mind the target demographic of the young and horny, there will be lots and lots – *and lots* – of sex: 'Celebrity Suck-off Special' will give has-been – or never-really-was – reality TV show contestants the opportunity to be the stars they always wanted to be; 'Dwarf Orgy' is what exactly it sounds; 'Simulated Rape – Live' will involve porn actors playing out male and female viewers' fantasies, which will be sent in by email or mobile phone in real time. There is also the opportunity for all X-TV viewers to upload their own films – taken by mobile phones, for example – and these will be available to the public to download and enjoy at their leisure. Anybody and everybody will be able to become a star on X-TV. It is user-generated content to the max!

Rasmus knows that people will love X-TV's New Season, that these shows will give kids who dream of nothing else the opportunity to be famous – which is all most of them want from life anyway. One or two of them may even rise above reality TV celebrity to become stars and household names. And the viewers at home will absolutely love it all, because they will see that people have become famous by just going on some reality TV show. And that means that no matter how useless, talentless, lazy, ignorant, ugly and stupid they themselves are, they know they can become famous one day too.

Of course, the viewers will also hate those made famous by these reality shows, as they always do. It is the arc of fame that is always followed, in which devotion and love for the famous becomes envious bitter resentful hatred – a trajectory often seen but usually ignored. This applies even more so to those who do not deserve fame at all: the talentless, unachieving wastes of space whose celebrity has been spawned on reality TV.

The contestants on these shows always, but always, live under the misapprehension that the public actually love them and are interested in their sad little lives and their half-formed dumb opinions. They just never get it. They never ever realise that it is they who are always the joke, and that everyone is laughing *at* them – not *with* them – for the desperate fame-hungry attention-seeking freaks that they are. Of course, most will be forgotten within weeks, months or a year at most. Like the dead.

'Thank you for your attention, people,' Rasmus says to nods and smiles and thanks. 'I've just got one more piece of news.'

Thursday smirks at this point and bites his lip – a rare slip of self-control for him. He always loves the way people can be so easily bought. It is the same all over the world, and it is always so easy. It reminds him of the cockroaches in the villages in his homeland – how he and his comrades took their gold and banknotes and bodies, and promised their lives would be spared, only to cut them to pieces like bush meat. You get used to the screams and blood after a while – learn that death is just another part of life. So why not learn to enjoy it?

'Everyone gets a bonus of twenty five per cent on top of their salary with immediate effect, as a thank you for the great launch of X-TV.'

The X-TV producers are delighted and amazed. They are not used to being treated like this by television production companies – in fact, they are more used to those companies trying to haggle their fee for freelance work down to bargain levels. Their salaries are now double what any of them has ever earned anywhere else.

*

That evening, Rasmus Karn – in common with a large and growing portion of the population – watches X-TV at home. This is a rare event for Rasmus, who, like so many in charge of television, watches remarkably little of it himself.

Instead, he listens to music – classical music, and Tchaikovsky, in particular – and does that thing that, according to our media culture, is a weird, unpopular and pointless thing to do: he reads books. Novels by Dickens, Defoe and Swift and more, from centuries past, plus poetry and plays including the classics from ancient Greece and Rome, and non-fiction books on history and science and philosophy.

All he wants to know is there – in these books – and reading them gives him knowledge, pleasure and power. TV has nothing at all to offer him – not as a viewer, anyway. No need for a screen if you have a book. For if books are the children of the brain, then TV is its slave master.

Thursday is there too. His role is to always be there – at work, at home, wherever Rasmus is. He knows that without Rasmus he would have no life, so he will always be there to defend and protect his saviour.

Home is a seventeen-roomed white stucco mansion in Holland Park. As on every day, Thursday has driven Rasmus there in the Mercedes, and made sure all housekeeping staff have left – they are only in the house when Rasmus is not, which is better for all concerned.

The house is decorated sparsely, in a similar manner to the office: there are few colours, except for the slim and richly-patina-ed antique Georgian furniture which stands in most rooms, and the occasional antique porcelain vase. The only pictures on the walls are prints by Gillray, and old shots of famous movie stars.

A huge flat-screen television hangs on the wall of the rectangular living room. It is the only TV in the house. The other rooms all have shelves lined with books, old and leather-bound as well as newer tomes, and a quick perusal of these would show, from the creases in the spines, that most have actually been read. Other than the television, all other technology is hidden away behind black wooden panels – the CD and DVD players, and all the computer equipment, laptops, notebooks and satellite equipment.

Rasmus sips tea from a bone china cup and watches the screen – as millions of others are watching their screens at that very moment, all over the UK and the world, in anticipation of what is to come. For on that evening, 'Suicide Bomber Week' is reaching its first, bloody climax.

Suicide Bomber Week - Live

The boy is young, perhaps eighteen, and small for his age. He is seen on camera leaving the back of a van in some dusty village is Afghanistan.

The camera is far-off but the zoom is excellent and you can even see the little wet beads of sweat glisten on the boy's forehead.

Beyond where he stands, scared and bewildered, there are several men at some sort of make-shift cafe, drinking tea or coffee or whatever Taliban fighters drink. They have guns, but they are not holding them: they are mostly propped up nearby. Only a couple of Taliban have them slung over their shoulders within trigger-finger reach. The men look relaxed. They are at home here, and safe. They know they are in charge – that this is their country now and always will be.

On the road – in reality, just a dust track through the village – people are ambling along and pushing carts. It has been a busy market day today, and some stalls of vegetables and fruit are still laid out. Old men and veiled black-burka-ed women barter and haggle, young children annoy their mothers and siblings, and life goes on.

The boy does not look out of place. He is just like one of the many teenage boys the Taliban elders take for themselves, to train for martyrdom, and other, more intimate duties and sacrifices. So none of the men with guns takes much notice of him walking along the dirt track towards them, not until they see the terror and sadness in his eyes – that familiar look of silent futility, of fear, of acceptance, of fate – and then it is too late.

The boy presses the button; the charge is released. The end of everything is here and now.

A blast of burst flesh sprays blood-splatter onto the walls and sand. The sight of the world exploding precedes its sound on the TV screen by over a second. Dust everywhere, an exploding sandstorm. Cries and wails in the shock wave. The *cack-cack-cack* of gunshots. Shouts. Cries. Shooting. *Rat-tat-tat-tat-tat. Yack-yack-yack.* The air clearing to show wreckage – of everything. A devastated world, crunched up and broken like a child's toy. Puddled blood on the dirt track. Body parts – arms, legs, other things. A jaw bone here; a ribcage there. Death. Dying.

The men with guns are there too, dead or alive, writhing in the dirt, skin flayed and shredded by shrapnel from the blast. Some who are still conscious have managed to find their guns and are spraying bullets into the air, as they would at weddings in happier times, and trying to see a target who no longer exists because most of his body has already been vaporised into molecules of blood and soaked into the sand. Some of these Taliban fighters have no hands or arms left to shoot, but at least they are alive – there is practically nothing left of some of the others. Those who are injured and limbless try to stand up and fight on, brave like on the battlefield, welcoming martyrdom and their longed-for paradise, and running into the bullets shot by their comrades in the confusion. Friendly fire. It is Allah's will.

Coughing and crying and wailing rise up from the devastated scene like prayers or the voices of the martyred dead, though the burst eardrums of those near the blast mean that most are not heard. There is no trace of the suicide bomber boy, whose last memory was of the huge widescreen TV he would buy with his winnings if he survived. The camera zooms in on what could be his severed head, blown clean off in the blast, like a cork from a bottle – though it looks more like a coconut sitting in the sand, crimson with the blood.

Shame it's looking the wrong way, thinks the TV cameraman, whose primary concern now is to get out of there (and his distant hiding place) alive – to get the full face would've made a great shot.

'Yeah cool right yeah wicked!' says Danny Mambo on the huge TV screen at the outdoor event somewhere in London. 'Dat boy is history, man!'

The crowd cheers and whoops, as Danny and Nazish high-five in front of a manic mob, each member of which feels special that they have been there to witness TV history being made, and each of whom waves and jumps frantically at the cameras, knowing that their ambition of being on TV has finally been achieved that night. Several crowd members faint.

Some girls wet themselves with horny excitement – and more than one mistakes this for her first orgasm.

Exactly two minutes later, the scene is repeated: a girl this time, in another Afghan village which looks as desolate and dusty as the first. The girl does not know the boy is already dead – she still thinks that she has a fifty-fifty chance of being blown to pieces. This is what makes watching her going through with it – watching her agonising wait – even more fun, and keeps people watching. You can even see her trembling in fear under her burka.

She walks towards some Taliban men dozing at the side of the dirt track road and presses the button – quickly, nervously, with eyes clenched shut.

This time, there is no explosion. The girl wobbles slightly, as if she is about to collapse, blown over by an invisible wind, but just manages to stumble back to the van which brought her to this terrible place. As soon as the doors close, she rips the covering off her face and vomits. Then she starts giggling uncontrollably as she is driven away to start a new life.

It would really have been better the other way round, think the production team when they see this – if the first boy's detonator had been the one wired so as not to go off. Then the people could have seen the girl walk to her certain death over the dust track, and her suffering would have had a real climax instead of fizzling out in her thankful relief. Dead girls make better TV than dead boys, as everyone in the business knows. But no matter. The real-time ratings are excellent – beating the BBC yet again – and there are several more nights of 'Suicide Bomber Week' to go. Online forums frequented by teenagers are calling the show *'well sick'*, which is a definite compliment. Other online forums express outrage, shock and horror – which is also a compliment, as well as being the exact intention of the programme makers.

'Man, dat is well sick, d'ya get me?!' says Nazish, his big afro haircut wobbling like a wig – though he always swears it's natural.

'Well phat, bruv – wicked!'

'An' comin' up on X-TV...' yells Nazish.

'X-TV!' echoes Danny, booming like a rapper.

'We got da world's first...'

'An' da world's only...'

'Granny Gang Bang,' they chorus, to whoops and cheers from the crowd.

'So bring it on, Alicia, baby...'

'An' not forgettin' da main man, da cool Calvin Snow. Wicked!'

Granny Gang Bang

'Thank you, Danny and Nazish...' says Alicia, 'and welcome everyone to...'

'Granny Gang Bang,' choruses Calvin in his angelic singing voice.

'Bring it on!'

Calvin can't really believe that he's doing this – presenting a show where some 'lucky guys' get the chance to have sex with old women, live on TV.

He thinks back to when he went on PopStarzzz. He had dreams of making albums, being in the charts, going on tour, maybe even breaking America – like Ed Sheeran or One Direction. Then he was chucked off for being underage, which he and public opinion thought was, like, so unfair, though he had to admit he did lie about his age on the application form.

For a couple of rollercoaster weeks, Calvin Snow was watched by millions on Saturday night TV and was really famous, but then he was just famous for being the kid that the judges asked to leave for lying. And then – nothing happened. Calvin was suddenly famous simply for being a failure on a TV talent show, and more than that – he was completely broke. People don't realise that just because you're famous and have been on the telly, it doesn't necessarily mean that you're rich.

The only difference between after and before he became famous was that after he got called names and laughed at in the street – and, on one occasion, was even kicked and punched to the ground by some other lads his age after he couldn't give them the money they demanded for the simple reason that at the time he only had 27p in his pocket – which was the whole reason he was walking home and not getting a bus or a taxi. They're just jealous, he thought, as they kicked him in the ribs again, chanting his name like the PopStarzzz audience used to: *Cal-vin Cal-vin Cal-vin*. How could he forget that?

After PopStarzzz, he was offered an advert or two, and a magazine feature, which brought in a small amount of cash – apparently, they only pay big bucks to well-known celebrities, not those kicked off TV talent contests for lying. But that was it, apart from an offer for a hardcore gay porno movie which he'd *never* do. No way. Not even straight porno. So what was he to do? He wasn't academic, so for someone like him, the only jobs going were in the local chicken-packing factory – which is perhaps why most people he'd gone to school with were now on the dole or in prison.

But people said he was good-looking, and not just his mum either, so why shouldn't he use that to make a life for himself? So, much as he hated himself for it, how on earth could he refuse the offer from X-TV to audition?

He only hoped his mum and grandma weren't watching 'Granny Gang Bang', though he knew neighbours in Sheffield would tell them what he was up to. Calvin had only told his family he had a TV presenter job, not this – not a sex show. But what else could he do? He had accepted that the X-TV presenting job could mean any kind of TV presenting, and he was just delighted to be back working, and on telly too. It was a start and anything was better than the chicken-packing factory. It could lead to something else, something different – something decent and respectable that would make his mum proud.

Anyway, lots of famous people started by doing something dodgy – Marilyn Monroe did nude modelling and Madonna did a porno flick. And film stars were all hired for their looks really, when you thought about it: if Johnny Depp or Brad Pitt were all fat and ugly and covered in zits, then they wouldn't be movie stars, would they?

And it wasn't as though Calvin himself was *"getting his kecks off"*, as he always said – he was just presenting a TV show where other lads did that. This thought was worming its way through his mind as Alicia introduced him:

'So big it up for X-TV's cute new presenter, Calvin Snow,' screams Alicia.

Girls shriek and scream, and some start chanting his name.

'Cal-vin! Cal-vin! Cal-vin!'

Calvin smiles that perfect dimpled cherub smile and blushes beautifully at the camera – the producers know that it is the innocence of this boy, his total lack of awareness of his angelic looks, which will make him so irresistibly attractive to TV audiences.

'Thanks...wow...hiya everyone...y'alright?'

More whooping and screeching from the girls, and not a few boys, in the West London X-TV studio where the show is being filmed. Calvin's little boy blue eyes blink winsome innocence into the camera. He feels stupid, as he always does when on TV, but his bashful blushing expression makes him look even more attractive. Then he remembers the script:

'Tonight, we're gonna be showing you something never seen before, like, on't telly...'

'And boy, do we mean that!' says Alicia.

'Cool,' says Calvin, not sure what else to say. 'This is reet good, like!'

Alicia looks at him with a fixed smile. She does not know why Calvin keeps on saying that, and in that weird accent too, which she thinks sounds like it's from somewhere northern and foreign – Scotland, perhaps, or Newcastle.

Screams, shrieks and squeals – the noise of sex and fame.

Before long, the TV screen is alive with bodies, young and old, frantically fucking in an orgy of bad taste and buggery. The people love it. 'Granny Gang Bang' is a big hit, live and online.

Calvin Snow is famous again. And he feels so ashamed.

Reaction

The ratings put X-TV at number one – which means their advertising revenue is rising exponentially too. The money is flooding in – which means more such programmes will be able to be made, which is exactly what the people seem to want.

But there is outrage in the media – not because of any needless deaths of innocents in Afghanistan, but because of the shock of explicit images of intergenerational sex in 'Granny Gang Bang'.

Minty sees the public reaction and the press coverage. So does Rasmus. He knows that a strategic response is needed and he has to do something, for the sake of good public relations. Without public approval, X-TV would cease to exist, and reputations can be lost so quickly in the media world, as can advertising revenue. Rasmus is not stupid. Contingency plans have been worked out well in advance. A reaction of shock and outrage has long been expected, and how to counteract this has been discussed and planned for – strategies are in place to cope with, and counteract, every outcome.

And so it is that, even while politicians of all parties and media commentators are protesting and pontificating about lowest common denominators, taste and decency, 'civilised' values, and the need for internet regulation, X-TV is simultaneously holding a press conference. Rasmus himself announces that all proceeds from 'Suicide Bomber Week' will be going to help injured and disabled servicemen and women, and their families, as well as the children of Afghanistan who all too often have lost limbs to bombings and landmines, old and new.

It is a stroke of media genius. Rasmus knows that no-one can possibly criticise X-TV now, not when they are giving so generously to charity, not when the Afghan authorities as well as Armed Service Charities and the families of dead and injured service men and women have given their approval. There is even an invitation to Prince Harry to accept the cheque on behalf of the service charities. It will be into seven figures, though it will only include the revenues earned by the phone poll and direct advertising, and not the vast income X-TV has made, and will make, from the gathering of information about viewers, data that can be mined for future commercial benefit.

Minty watches all this unfold on TV that evening. She knows this is a brilliant bit of manoeuvring by X-TV and Rasmus Karn, and wishes she'd thought of it herself – though the BBC would never allow so much violence, unfortunately, and couldn't give its money to charity anyway, due to the inconvenient fact that it belongs to the public. She knows the BBC is in crisis, however. It is haemorrhaging viewers, especially the young, that key under twenty-five demographic, and losing them to X-TV. She knows too that this has to stop, and she knows there is only one way to do that – by hiring Peter Baztanza and making the BBC a serious competitor to X-TV and others, especially online, with new, exciting and ratings-grabbing reality shows.

But this is only one part of her strategy. The other is just as important – she must find out who Rasmus Karn is and how she knows him. She knows that this is the key to destroying X-TV, so she has asked her long-time assistant, Toby Tickell, to compile a list of all people, with photographs if available, she has had contact with at the BBC – in meetings, in the commissioning and producing process – over the previous ten years. The list will stretch into the thousands, but it is the only way for her to find her man.

It is only by cutting off the head that one can kill the snake. So Minty will stop at nothing to achieve her aim – to destroy X-TV by destroying Rasmus Karn, whoever he may be.

When she thinks more about it, the rise of X-TV may well not be such a bad thing, after all. It may well even have given Minty Chisum the opportunity she was looking for to make her dream a reality – to rise to the top of the BBC and take charge at last.

Hack

X-TV has access to vast amounts of information about its viewers' behaviour. Some of this data it gathers itself, using experts the company has hired; the rest, it receives from various agencies.

Information is harvested every time an individual uses a mobile or goes online, and nothing can be done legally to stop these personal details being accessed. X-TV has full access to this information about its viewers and uses it to full effect.

Despite what the internet companies, financial institutions and national governments may say, no digital information is truly secure – ever. It can all be accessed by those who know how, and accessed in such a way that those whose duty it is to store the information will have no idea there has been a security breach.

Rasmus knows that the BBC will be trying to find out as much as they can about X-TV, and about him personally too. He also knows that Minty Chisum will recognise him, however much he has changed since he knew her, and that sooner or later she will realise who he is. This is to be expected, and Rasmus is prepared.

Minty does not realise it, but her movements and messages have all been available to Rasmus for weeks. And it has all been done legally too – no phone calls or messages have been illegally hacked. Rasmus and X-TV are always careful to obey the law – to the letter – according to which jurisdiction they are operating in.

People really have no idea how much they give away about themselves, and they have absolutely no idea just how much they are being watched or who is doing the watching.

It is through X-TV's IT monitors that Rasmus is aware that all X-TV mobile phones and smartphones are being hacked. It is no surprise – in fact, it has long been expected. The only real question is *when* and not *if* this would happen. X-TV staff have always been ordered not to use mobile phones when discussing anything confidential. But then, only Rasmus – together with Thursday – truly knows the strategy of X-TV and where it is heading. Moreover, the X-TV offices are fully protected by blocking devices which prevent anyone using remote spying devices to try and listen in to conversations within the building.

Rasmus knows Minty will stop at nothing to get what she wants, and also that she has the intelligence and determination to get it. The information X-TV now has about her means that at any time they could report the hacking of their phones to the BBC or the police, if they chose to do so.

But for now, they will do nothing but just watch the watchers and their growing desperation. Rasmus is enjoying himself.

Blog

Rasmus blinks into the blue electric light of a computer screen.

He thinks, he smiles, he types.

He knows his words, though not live online, will be valued in future times when what he is doing will all make sense, when all the world has changed. History demands that he does this, that he charts the course of this revolution.

Minty is there, in his head – there, just a mile or two across West London from Holland Park.

She sits at her laptop too, searching for a name, a face, anything.

She knows she has met Rasmus before, but just cannot find him. It's as if he has melted like a ghost into the mist of her memory, as if the more she grasps at the image, the more it blurs and fades like smoke.

Rasmus types history into the computer screen. One day, Minty will read this blog and know the truth, know the full significance of X-TV, know that what will happen is, and always has been, inevitable and definite as death.

His hands type the words:

I am Rasmus, and this is my reality.

The blue electric light from the screen glows and flows like liquid lightning through the room, glints glassy in the eyes of Rasmus who reaches out and turns off the computer, leaving the screen to hum itself into its cold still darkness.

Shut down.

Numbia

Rasmus is video-conferencing with Mercy and Gary Wu, who are now on the ground in Southern Africa with their OB (Outside Broadcast) crew. The programmes they have been preparing for many weeks and months are nearing their conclusion. 'Live Shark Attack' will, they know, be a killer show.

'Yah,' says Mercy on the video link, 'all is good here.'

'Too flippin' hot though, innit?' says Gary Wu, wiping his eyebrows.

'You Chinamen just too cold, yah,' says Mercy, 'not like us black folk with our hot, spicy blood!'

They laugh; Rasmus smiles; Thursday grins. Mercy reminds him of a woman he raped in the war – several women in fact. He knows what Mercy will look like with her belly cut open and her breasts cut off. He knows what her brains would taste like too. He keeps his thoughts to himself.

'Thursday, sweetie – you know I don't lie, yah?' giggles Mercy.

Thursday laughs a great big wide laugh, deep bellowing guffaws ejaculating from deep within him, which is all instantly converted and compressed into the encrypted binary code of digital communication, beamed up from London to an orbiting satellite miles above them, then unencrypted and picked up by the laptop in a hotel room in Africa. There is still a slight delay, and occasional picture distortion, but this does not seriously hamper communication.

'Stuff's sorted over here, innit,' says Gary Wu. 'All the contestants are in the hotel, mooching about...'

'Just waiting for mister shark, yah!' says Mercy. Her affection of saying 'yah' at the end of most sentences is something she acquired at Cheltenham Ladies' College, and not, as many may think, part of her black African heritage.

'Shark Attack – Live,' as Gary would say, 'does exactly what it says on the tin, innit?'

That is why they are in Numbia, a small free state in southern Africa, ruled over by the benign dictator President Bombobo, whose tolerance of those who criticise him is expressed by generously allowing them quick execution with a bullet, a noose or a slit throat, rather than the prolonged agonies of torture, which they would suffer if they didn't confess to their crimes of treason immediately. He is also benevolent enough not to slaughter the families of most of those who are executed, unless the traitor has pleaded their innocence, of course – then they have left the President no choice by their selfish and hurtful behaviour, which is all most tragic and upsets the President terribly.

President Bombobo is a good friend of both the United States, who support his free market economic policies and good Christian values (which have involved massacring non-Christians in the past), and China, to whom he sells off Numbia's abundant natural resources for vast sums, the majority of which end up in his own family's foreign bank accounts.

Numbia has the advantage of having some of the most beautiful coastline in southern Africa, and what is believed to be the largest population of great white sharks in the world. The sharks are now hungry, ready and waiting – just like the contestants who, very soon, will provide their lunch.

'Thank you both for going out there to produce this broadcast,' says Rasmus on the video link. 'I know it can't be easy in the heat.'

Gary wants to scream his hatred of the heat at the webcam, but both he and Mercy simply say how much they're enjoying it over there, despite the stifling temperatures, the power cuts, the dirty-looking water supply, the thuggish soldiers standing everywhere and, more than anything, the disgusting stench of open sewers, dead dogs and decay that hangs over the place. Last night, he even dreamt that he was breathing blood.

The video conferencing ends. Rasmus and Thursday will all watch the show live as it happens, together with millions online. The X-TV shows playing at the moment are dominating the schedules anyway, and 'Shark Attack – Live' will join them at the zenith of British broadcasting.

Thursday is still grinning. The pictures have reminded him of Africa – of the sweat, the corruption, the smell. The horror. He feels a momentary pang of homesickness – nostalgia, even – but the mood soon passes, as all things must.

Rasmus observes Thursday's face. He can see the yearning in it, the hunger for something more.

'Patience,' says Rasmus. Thursday smiles at his saviour and nods.

Names and Faces

Names names names. Minty sees them all – all those morons and fuckwits and retards she'd met at the BBC; all the no-hopers and talentless bullshitters; all the writers and producers and commissioners and the rest. Name after name after name from her past, at the BBC. List after list of them, page after page after page – each one a link in a chain weighing her down.

She tries to imagine a face for each name – which, most of the time, though not always, she can do. Then she looks at the file Toby has prepared for her, of both names and faces, to see if there is a photo available to match the name. With this process she had narrowed the field down to thirty, plus another twenty uncertain, because there are no photos of these people in the file.

It is only through such meticulous work that one's targets are achieved, Minty knows, and she spends over seven full hours going through the list. So engrossed has she been that she has not noticed how the afternoon has merged into evening and it is now after 7pm. She loves moments like this: moments marinated in the kind of focus that has got her where she is – the deep creative *flow* of concentration in 'the zone' which makes time stand still in the mind and which can lead to great creative achievement.

Minty squints at the names and photos at her desk. She opens a new, blank notebook and writes each name on the top of a separate page: she knows that things always seem much clearer when written down, far more than on a computer screen, and she knows she thinks this way because she is not young. Then she takes the notebook and pen, walks over to the sofa, lies down and closes her eyes. In this way, she will be able to conjure up an image in her mind of each and every person, if she concentrates hard enough. That is the plan, anyway.

She is close now and she knows it. She remains convinced that Rasmus is one of the people on her list, though his name must have changed since she met him. Maybe he even has a photo in the file, though there is no great similarity between any portrait in there and the photo of Rasmus that she has seen. There is no-one with a scar on his forehead, anyway, so either there is no photo of him or he has changed and acquired that scar since the BBC photo was taken.

All she now has to do is lie back and think, think and think again – and hope that a face and name will materialise in her mind.

Baztanza

'But we can't pay that much – you know that, Peter,' says Minty, her Tourette's having mysteriously vanished in Baztanza's presence.

'My fee is my fee,' says Sir Peter. 'Fantastic! I'm loving this!'

Minty tries not to sneer. She hates the way Baztanza's wonky grin seems to take up most of his face almost as much as she hates kowtowing to him, but she also knows that there is no other way.

Sir Peter Baztanza is having a meeting with Minty Chisum, a woman he has met briefly before and who, he knows, is now Head of Entertainment at the BBC, no less – so she is in fact his direct and state-subsidised competition. They have spoken on the phone earlier, and she has mentioned money – and lots of it too.

Baztanza has never knowingly refused to meet anybody when there is a possibility for him to make more money, despite the fact that he's never really needed it. His paternal great-grandfather Archibald Baztanza founded the most successful manufacturer of lavatories in 19th century Britain thanks to Queen Victoria's patronage, which made him, quite literally, a shit-load of money.

Sir Peter Baztanza is pacing back and forth in his office in central London. His gait is that of a young fit animal, a leopard perhaps, as he prowls the room bouncing on his springy heels, primed to pounce on any passing prey. His expression is one of permanent wide-eyed surprise which, together with his always-spiky hair, makes his face resemble that of a child hanging upside down on a climbing frame.

'That's why we need you,' says Minty, deciding to be upfront, polite and honest - which she herself would acknowledge is a rare state of affairs for her.

Baztanza grins at her, grinnily. He knows that Minty - the personification of 'Auntie Beeb' today - must be desperate indeed to come grovelling to him, because, as he well knows, the BBC never knowingly spends a pound when a penny will do.

'Fantastic!' he says, pausing momentarily to enjoy a big-breasted girl in a bikini get covered in green slime on the game show showing on one of the giant TV screens on the wall. Peter Baztanza has a great big grin on his face - not a smile, a grin - and it is a grin which is always there, so much so that it has become his brand and his trademark. His TV production company is even called 'Big Grin Productions' - its logo is a great big grin too.

'OK, Minty,' says Peter. 'I'll tell you what I'll do. I can give you a little discount on my fees - say, ten per cent?'

'Thank you,' says Minty, perhaps ironically.

'In return for an increase in my percentage of future royalties, of course - to 45%.'

Minty doesn't know what to say.

'OK, Peter, I'll ask the DG,' is what she actually says.

'Fantastic!' says Peter Baztanza. 'I'm loving it!' And he taps the remote to change the channel of one of the TV screens. The big-breasted girl covered in green slime now seems to be minus her bikini.

'I'm loving this!' he says, when Minty has left, a widescreen smile stretched tight across his face.

Wasps

Hugo stands on a hillside and breathes the fresh Spring air - not ordinary breathing, like in the city, but deep lung-bursting countryside breathing, organic and natural and good-for-you. In and out, and in and out, and in and - Oh! The glory of it all!

Bliss it was that day to be alive – right here, right now, breathing in the English air well away from the traffic fumes of London! He coughs up car-exhaust-coloured phlegm and spits it out onto the green, lush, life-affirming grass.

If Hugo's eyes were better, he would be able to see from the hillside where he stands to the little village below – a traditional English village, full of traditional English craft shops, estate agents, take-aways and a gastro-pub, past the church and its steeple, which has been gradually falling down since 1964 – across to the recently-closed primary school, to the land beyond with its NO TRESPASSING signs on its high wire fences where a brand new Tesco megastore and its car park will serve its customers, some of whom will soon be living in the two hundred luxury executive homes due to be built nearby.

Oh! The glory of it all! As Hugo would no doubt say if he could clearly see the world stretching out before him.

But Hugo can't see anything much, because Hugo's eyes are next to useless, even when he's wearing his thick little glasses. This is why Hugo has not seen that, instead of being beautifully bucolic and balmy and full of glory, the countryside all around him is a seething mass of disease, decay and death. If he'd been able to see properly, he would also have seen the unusually large number of wasps which were flying around him, attracted by the sweet stench of death from a wood where a farmer has dumped a sackful of poisoned and decomposing rats.

Hugo looks, but sees none of this, though he does hear a strange buzzing which he assumes is just the bees out and about being bees, off to pollinate pretty flowers and in doing so collect the nectar to make all that yummy organic honey. So Hugo keeps breathing: in and out, and in and out, and in – goes a wasp, which stings him once (on the tongue), twice (on the tonsil), and thrice (on his upper lip as Hugo spits it out).

'Bloody, um, wasp!' says Hugo – or, rather, *'Bubbie om boss'*, which is all he can manage with the pain – although he feels instinctive sympathy for the poor little insect on the grass which, Hugo notices, he has accidentally bitten in half. Its head is still going through the motions – it can see its attacker and keeps trying, with futile perseverance, to fly up to sting him again. A pang of guilt stabs Hugo in his stomach and he starts to feel sick.

'Poor um little thing,' says Hugo – which comes out as *'Boor om bipple fwing'*. At that very moment, another wasp stings his ankle, and another his wrist, and another his left buttock. It is as if his whole body has been stung in a synchronised wasp attack.

'Ow!..um...Ow!...um...Ow!' says Hugo, deciding to run down the hillside and towards the village he knows is nestling there somewhere, amongst the blue blur of the happy hills and vales.

'Bloody unlucky,' thinks Hugo as he stumbles along holding his jaw. It is only because he accidentally entered the wrong postcode into the SatNav that he's out here in the first place. He had meant to go to a BBC health and safety training meeting in Borehamwood, which is why he's dressed in his usual formal suit and tie. Now he has no idea where he is, but he knows he's bound to find some friendly local in the village with an effective herbal remedy for the really very painful pain he's feeling right now.

'Only a wasp being a wasp,' thinks Hugo, tolerantly, 'and wasps can be very useful to the food chain and in eating parasites and bugs which may otherwise damage plants'.

But the wasps are not leaving him alone. In fact, they seem to be following him – a little yellow-and-black cloud of anti-social insects buzzing after him in a quest to sink their nasty little stings into his skin. Hugo runs faster and faster, tripping and stumbling down the hillside, howling to himself.

And then he says it:

'Buggin, om, baabaas!' says Hugo – which, if anyone had been standing close by and had been able to decipher the words emerging from his swollen mouth, would have understood as 'Fucking, um, bastards!'

Even fans of organic gardening have their limits.

'Sorry, Minty, but we just can't do it.'

Minty has had enough. She stands up to leave the room, mouthing abuse under her breath as she slams the door behind her. The meeting with the Director General has not gone well.

'Fuck this for a game of cocks and cunts!' she barks at the timid-looking mousy secretary in the foyer.

So this is the way it's gonna be, Minty thinks. Same old BBC, mired in deep conservatism and risk-aversion, years behind everyone else and unwilling or unable to change. Same old same old. Continue paying peanuts and continue getting programmes made by monkeys' arses. You get what you pay for – no more, and no less. At least her upbringing taught her that, if nothing else – where she came from you had to be creative and take risks in order to get out, in order to survive.

According to DG Benjamin Cohen-Lewis, Sir Peter Baztanza was 'not really a BBC person, not one of us, in his attitudes and values.' He would be a 'mis-match', and an expensive one at that, no matter what ratings his shows got for other channels.

Benjamin Cohen-Lewis – calm, collected, managerial, number-crunching, phlegmatic Ben Cohen-Lewis – felt a tiny twinge of anger stirring in his usually dull and docile mooby breast.

'But Minty,' he said, 'there is more to making television than getting high ratings, though that's all good too.'

Minty closed her eyes and groaned agape into her very soul.

'I hear what you say, Minty, I really do, and it's all good,' continued the Director General, 'but, apart from anything else, Peter Baztanza is just too expensive – almost a million a show? Plus 45% of revenue? I'll never be able to get that past the BBC Trust, not at this present moment in time.'

'Fuck the BBC Trust,' Minty said, shocking even herself,

condemning the twelve mysteriously-chosen individuals of the BBC Trust who have power indeed, despite most of them never having worked in TV or radio at all.

'The BBC gets rather a lot of...er...ratings too, you know,' insisted the Director General, 'and...'

'We need Baztanza,' insisted Minty. 'It's the only fucking way for the BBC to make great programmes again!'

'Actually, Minty, we've got a new and exciting diverse range of programmes in production right now,' said the DG with pride, 'and we've got some wonderful new producers, directors, writers and presenters – for example, Kevin Kumar, Gbemisola Eniola, Wendy Wang, Femi Umunna, Kweku O'Reilly, Tabitha Twistleton, Beany Cummings, Miranda Mupp-Spoon, Caspian Weeble, Clarissa Calypso...'

Names names names.

Minty closed her eyes and listened to a roll call of the BBC's new flagship 'talent' – all the well-connected, croney-promoted, vibrant and diverse producers and presenters who would take the BBC into a bright and excitingly new future.

'...Stella Snowball, Violet Flange, Gavin Puddle, Sergei Slipenchuk, Kok Yong Phat, Diana Drummond-Jones, Benidorm Bandersnatch, Vince Crisp, Maryan Siddiqui, Jerker Maelstrom, Tomi Wallop, Della Beach-Spode...'

And still they kept coming: the names, the names, the names – the adenoidal roll-call relentlessly dribbling like snot from the spreadsheet-stuffed skull of Benjamin Cohen-Lewis.

'...Mariella Montague, Jasper de Quincy-Monckton, Vinod Zardari, Ptolemy Pepper, Cheng-Chu Chin, Chin-Chan Chong, Nigel Noote, Otto Koo, Gareth Grampus, Ted Head, Leslie O'Greedie-Wonkysmyle, Louis Clack, Gabby O'Kwerty, Morgan ap Merrick, Hermione Humm, Clitilda Clump – I could go on.'

Minty scowled at the DG, wearily. No, don't go on. *Please* don't go on!

'I'm particularly looking forward to Animal Crackers,' chirped the DG, 'which, as of course you know, Minty, was originally an idea to emerge from an exciting BBC thought shower workshop.'

A 'thought shower' was what used to be called a 'brainstorm' – but the new diverse BBC no longer used expressions that might possibly offend anyone with any medical condition – in this case epilepsy – so 'thought shower' was the accepted phrase in the inoffensive linguistic landscape of BBC-land these days.

'We British do love our animals, don't we?' says Ben Cohen-Lewis. 'When I was a boy, I had a goldfish and a hamster. Absolutely adored them, I did. Until they died...'

So now, instead of cutting edge programmes by the king of reality TV, Peter Baztanza – someone who could, albeit for a whopping fee, have really done the business – the BBC was going to make a talent show featuring pets in the vain hope that they will save it. And all because it wouldn't cough up a few million quid, out of its huge multi-billion budget, to hire the best.

'Cunt!' Minty shouts up at the DG's office as she walks to her car. She calls Peter Baztanza on her mobile and breaks the news to him. She also says that she will keep in touch, in case of developments. And she will, because there will be developments alright.

'Fantastic!' he says, as expected.

Peter Baztanza knows what Minty is up to. He can smell the ambition on her like week-old sweat. If Minty ever becomes Director General – (and he knows how desperate the diversity-worshipping BBC is for a woman to occupy that role) – he knows firstly that she will have real power, and secondly that she will be desperate to hire him, and for almost any price. Because if there is one thing Peter Baztanza knows it's his own worth.

The BBC, despite being so irritating, slow-moving, institutional and generally up-itself, is nevertheless a great national and international brand, and one which brings with it automatic customer recognition and respect worldwide –

something that makes it worth billions to those who know how to exploit the brand.

And Peter Baztanza is greatly looking forward to exploiting the BBC as much as possible, when the time comes.

*

Money money money. It's always all about money.

Without money, you cannot make great TV programmes. And without great TV programmes, you can't attract a large audience.

And without a large audience, you can't generate revenue from advertising and sponsorship to make more money.

The BBC likes to think it is above such vulgar things, that it exists in a kind of non-mercenary state within a state, as refined and aloof as a benevolent monarch or dictator, where no commercial pressures will – or should – ever penetrate its perfect world of patronage. It is always mightily annoyed and offended when commercial pressures are brought to bear on it, whether from the government, the media or the people.

In many ways, life at the BBC is like life in a communist state. Its rulers and administrators are arguably as out of touch as any Soviet Commissar, unaware of anything and unable or unwilling to believe that anything could possibly go wrong, because it is written that the system was designed so perfectly that it can never fail, and can never be improved either.

Indeed, the BBC is the nearest thing to Communism that the British have ever known.

The fact is that there is plenty of money in the world, and it can be accessed by anyone, if you know how. And those who control it, control the world.

The politicians and bureaucrats who give the impression of being in charge are merely posturing puppets – they are not in control of anything at all. Not really.

It is those people who control the world's money who

control the world's media, including the internet, and so they control the world's people too. That is real power – a power politicians no longer have.

It is a power that is now held by the new Masters of the Universe: X-TV, Google, Facebook, Twitter, Youtube, Apple, Microsoft and others – the new and better rulers of a new and better world.

I am Rasmus, and this is my reality.

*

Email - Marked URGENT!
From: Hugo Seymour-Smiles
To: Minty Chisum; Oliver Allcock; Lucinda Lott-Owen; Sangeeta Sacranie-Patel
Cc: Benjamin Cohen-Lewis, Penelope Plunch
Subject: New Meeting

Hello everyone.

Just a quick email to remind everyone about this week's task force meeting.

Unfortunately, due to a wasp-facilitated injury resulting in lingual tumescence and an acute malfunction of the oral cavity, I am completely unable to speak at this present moment in time, so will not be attending the meeting, going forward.

Oliver Allcock, as Controller of BBC One, will be standing in for me.

I shall, however, be on the ground at the exciting live broadcast of "Animal Crackers", so will be in the vicinity if you want to talk about anything, going forward, though I will not be able to talk for lengthy periods of time, due to aforementioned insect-induced oral issue.

Please let both myself and Oliver know if you cannot attend the meeting.

Regards,

Hugo

BBC Head of Vision

*

Live Shark Attack

It has been a sultry, scorching, oppressive afternoon in the Democratic Republic of Numbia.

Only now, in the early evening, does a balmy breeze blow through the sullen stillness of the air, sucking tiny droplets of moisture from it as a mosquito sucks blood. But the white sand of the beach is still scorching hot on the soles of the fearless yet frightened feet which stand upon it.

The contestants are mostly black African, though plenty of white Westerners, already rich as kings by local standards, have been unable to resist the lure of fame that has drawn them from the more peaceful and prosperous corners of the world to the Numbian coastline today.

There are over a hundred contestants on the beach, wearing tightly colourful swimming trunks and bikinis of various designs. The swimmers' faces peer out over the shimmering sea to the island which is to be their destination. Gentle little waves lap innocently on the sand.

From the beach, the outline of the island and its trees looks menacing and haunting, as if witches or monsters are skulking there, deep in the woods, waiting for their next victims.

The ocean itself ripples with tranquillity and benevolence, its surface only disturbed by several bobbing boats on the waves to the left and the right, waiting – just waiting. An occasional swish of bubbles rises from below and, here and there, the tip of a fin cuts above the skin of the sea. There are monsters here alright, ancient and awful monsters, somewhere silent under the gentle little waves and mirror-smooth blue lilt of the water.

There have always been sharks here, ever since the creatures evolved into their perfect and terrible torpedoed form right here, in these waters, millions of years ago. But there have been even more since President Bombobo decided to throw the corpses of his enemies into the sea here - though, in recent years, it has been the writhing, still-living bodies of the tortured and bloody, rather than their cold corpses, which have been dropped into the hot blue ocean. It was observed that live prey is what the sharks really loved. The President has always been deeply committed to the preservation of Numbia's rich and diverse coastal wildlife.

In recent weeks and days, the shark population has increased substantially, too, since the army began pouring huge tank-loads of chum - ground up flesh, blood and bone, both animal and human - into the dark waters at the same time every evening, an operation carried out in close co-operation with X-TV.

It is now feeding time.

The swimmers know that the world is watching, and if - just if - they are lucky enough to live through the next hour, then they will be rich beyond their wildest dreams, as well as famous all over the world.

The white contestants are in it for the money, yes, but also for the thrill of it all and the fame: television is always about the fame, really. One, a New Zealander, is joking with an Australian co-contestant, each ribbing the other about the deficiencies of their nations and their swimming prowess. The Australian says he's out-swum sharks before; the New Zealander says they were probably just babies...

Near these two, some Africans are praying, mumbling with eyes closed, bouncing up and down, making the sign of the cross and or looking skywards to a syncretic pantheon of gods ancient and modern who they hope will soon welcome them to paradise if they die that day. The gentle oscillating hum of a helicopter overhead jolts them back to reality and the reason why they are there. Cameras capture close-ups of the contestants, and images of the scene are beamed via the wide blue sky into millions of homes all over the world.

Several contestants are crying; some vomit; others pee themselves. All have terror in their eyes. And then the hooter sounds.

Water. Into the white water. All the swimmers run into the white wild water. Splash splash splash. Into the giant aquarium of the deep and endless ocean as fast as they can. They all know that the first swimmers into the sea will have an advantage – they will be ahead and so will be just that little bit closer to the island when others, the ones lagging behind, are taken by the first sharks, swirling the sea with pungent clouds of blood, thereby attracting more sharks from miles around.

Soon, the beach is devoid of people, except for the film crews.

The surface of the sea is churned white with splashing swimmers. There is shouting and crying and, strangely, the occasional laugh – like a summer holiday down the local lido. No-one has seen a shark yet.

But they are there alright. Down, deep down in the dark depths below, those torpedo-sleek killing machine monsters little changed since their giant prehistoric ancestors hunted and killed other long-extinct creatures in the same seas.

And then the TV cameras see it: a fin. Then another. And another. Slicing through the water like sharp black knives.

There are screams as swimmers see the fins, a look of horror swallowing their faces in its dark and unforgiving jaws. They try to swim fast, faster, as fast as they can. But the panic only interrupts the rhythm of their swimming and leaves them spluttering and flailing in the water, bobbing like bait. And then the attacks begin.

The first scream is high and feminine, though it comes from a fat man who is one of the swimmers trailing the pack.

The sea around the man bleeds red as he screams a final scream through the blood bubbling in his lungs – a scream cut off in mid-cry as the shark drags him down. Then another scream and more blood gushing into the water. Soon, swimmers are being attacked on all sides.

The sea is soon red as an African sunset. Bits of bodies bob on its surface – innards and intestines and other things. There is a man who looks as though he is treading water, doggy-paddling: it is the New Zealander who was in the lead. A woman with terror in her eyes swims past him, nudging him slightly. This is enough to make his body roll over and up-end, showing where it has been severed at the waist, intestines spilling out like big red worms. A shark rises from the deep and bites more of the man's stomach away. The black woman swims on, her eyes now closed. She has seen enough.

If her eyes were open she would see sharks and bodies on all sides of her – a feeding frenzy and a bloodbath. She would see the TV cameras on the boats around her too, capturing the shark attacks in high definition detail, with top-of-the-range digital sound. And if she were to look up, she would see a helicopter shooting wide-shot footage of the swimmers heading towards the island – little human heads which turn into splashes and then disappear, leaving little watery-red lilypad shapes on the surface of the sea.

But she keeps her eyes closed. All she can see is her children's faces as she swims and swims. And through it all, the contestants keep moving forward, little by little, towards the island, though a small number turn back, which proves a fatal mistake in every case.

On the island stands X-TV's Nazish Naidu, huge Afro head bobbing excitedly as he jumps up and down like a hyperactive, sugar-rushed child. He is clicking his fingers – an action which makes his copious and chunky gold jewellery jangle – and laughing loudly as he watches the TV monitor set up on the beach through his trademark over-sized sunglasses. When not looking at the monitor, he looks out to sea at the splashes of swimmers approaching the beach. But it all looks pretty boring in real life, he thinks. The picture from the helicopter and boats being shown on the monitor is so much clearer than what you can see in the real world, shows so much more detail and blood. You can even see the expression on people's faces, the sheer horrific fear in their eyes as they go under – and you can see the black expressionless eyes of the sharks too, which is well cool.

Anyway, he thinks, he can hear the muffled screams of swimmers being attacked by the sharks in real time coming from the sea, so what with that and the clear close-up TV pictures, Nazish thinks he must have the best of both worlds.

His cameraman – a local man called Wellington who has filmed whatever President Bombobo has asked him to film for the last five years as part of the official government propaganda unit – stands waiting for the moment when the swimmers will come ashore, if any do. Wellington is not convinced any will arrive at all though – he knows just how many sharks are out there in the deep, dark, dreadful sea.

But it is not his place to question why; it is his job just to take the pictures. So he will film while the strange English brown man interviews the survivors on live TV. And people all over the world will see the pictures that he, Wellington, son of a goatherd, has taken. And President Bombobo will be happy with what he has done, and bless him with a happy and prosperous life as his reward.

On the mainland shore, Danny Mambo looks out to the island:

'Yeah cool right yeah wicked!' he says, as he watches the waters churning with blood and bodies and shark fins. He forgets that he is not on TV any more – his presenting job having been done before the swimmers entered the water. He finds it hard sometimes to get back to normal after he's been presenting, as if the character he created for his TV career is now taking over the 'real' him completely.

Now, though, the action has shifted to the sea and the island, and there is not much for him to do, not on camera anyway – apart from doing the live link-up with Nazish on the island after he's interviewed the winners. Typical Nazish – gets all the credit for doing practically nothing. But then, as Danny Mambo thinks, what else do you expect from a fake black man who's actually a Pakistani?

And then they all see what happens to Nazish...

Mercy and Gary watch it all on their monitors. Gary looks nervous, but Mercy knows that they are safe in the hotel, under the protection of President Bombobo. Also, there are so many Chinese in Africa now under the protection of various despots and dictators that Gary Wu, with his half-Hong-Kong looks, will be perfectly safe. She knows too that Danny Mambo, a black man, will be safe on the mainland beach.

In London, a screaming crowd is shrieking and yelling at the carnage on a huge outdoor TV screen. Alicia McVicar and Calvin Snow are watching too.

Calvin is tired. He has spent every day of the last week presenting 'Dwarf Orgy', 'Granny Gang Bang' and 'Mystery Celebrity Blowjob' and all he wants to do is sleep. Strangely, though, the sight of the sharks and the blood, as well as the sheer adrenalin of live TV, has made him wake up. He now feels almost as buzzing as Alicia who, he notices, is fake-frightened, with her right hand held to her mock-shocked mouth, and her left hugging Calvin's waist as the sharks rip their victims to shreds in real time on the huge TV screens.

Calvin feels a swelling erection straining his pants, though he is not sure if it is there because of Alicia's hand on the small of his back or the sight of people being killed by sharks – perhaps both. Oddly, when he's presenting the sex shows, gawping goggle-eyed at all the shameless shagging in close-up on the giant TV screen, he never gets aroused. Just nauseous. Weird...

In the ocean, the Australian is still in the lead. His powerful front-crawl – the one his father taught him and the one that he has used almost every day of his life in the pools and seas of the lucky country – pushes him onwards towards the island and his destiny as a winner, a champion and a survivor. He is now no more than a few short metres from the beach. Soon he'll be able to stand up on the seashore and wade out of the sea. Nearly there! So very nearly there!

But then he makes a fatal mistake – he looks up, anticipating the warm soft sand of the beach, and the riches and fame that will surely be his. The smooth, lithe rhythm of his swimming is thus interrupted. That pause – that unmistakable signal of vulnerable struggling life – is all a shark needs.

The Australian feels the shark's smooth primaeval back brush against his thigh and he knows he is doomed, just like the tourist he saw once being attacked off Sydney – and there was nothing they could do then either as they waited for the strike. He can see the beach – he can hear the voices of the TV presenter there, see the cameraman filming him. He even can smell the heat and the greenery of the trees, almost taste their fruit. He is *so close* – he is almost there.

But he is not 'there' at all. He starts to swim again, as fast as he can. But he knows that jaws are rising beneath him. He knows he is doomed.

And then it hits. The great white teeth of *Carcharodon carcharias* clamp shut on his torso and cut him clean in half. The sound of cheering and clapping from the beach is the last sound the Australian hears before he sinks beneath the waves where his skull is crunched to mush in those powerful jaws as easily as an eggshell.

'Wicked, man!' says Nazish. 'D'ya get me! Rispek!'

Far, far away, in Australia, a proud father turns off a television and opens a bottle to mourn his only son.

But no-one else turns off. The whole world is seeing this – watching the winner stagger out of the waves. Around Africa there are cheers and songs and dancing – for the winner is black. Not some arrogant white mercenary chancer, but a proud African warrior!

It is a young woman, thin as sugar cane, who stumbles

onto the beach, with a stunned look of horror frozen on her face. Her name is Charity and she has entered this contest, despite knowing she will probably die, as a last desperate hope to get her and her two small children out of the poverty they were all born to. For some reason she is humming – (or is it groaning?) – as she stumbles onto the beach. She then falls onto her front, grabs handfuls of sand, and rubs it into her face, as if trying to cleanse herself of the blood of others that sticks in watery red stains all over her ebony body.

She sees a coloured man there, kneeling down, talking to her, but she cannot hear what he is saying. He does not even seem to be speaking proper English – and anyway, the woman's ears are whistling with the screams of what she has witnessed, and full of seawater and blood. She then starts sobbing and wailing like a baby and rolls into a ball on the sand, like a small black crab.

Typical, thinks Nazish: the winner's a nutter. How on earth's he supposed to interview a nutjob? Fortunately, Nazish sees another swimmer stagger ashore, and then another, then another. So, OK – the winner might well be a mentalist, but at least there are others to interview. In fact, there are over ten people on the beach now, all but one black, though a couple have limbs missing. The medics on the island rush to their aid, though they know there is no point, not after all that blood loss and the shock.

Nazish is doing a piece to camera and Wellington is doing his best to keep the presenter's huge Afro in shot – not something he has ever done, because no African he knows would ever wear such an unkempt and unrespectable comedy bubble hairstyle.

And then something happens. The people on the beach see and hear Nazish asking them questions, and smiling – actually smiling – at what has happened. So many people have just been killed and eaten alive by sharks, and this man – this *coloured* man – is smiling and laughing to camera?

It is a woman who acts first – a proud African woman who

has heard how the Asians used to treat blacks like her family in the old days, until the great President Bombobo dealt with them. She remembers how the coloureds took the wealth and resources that were theirs, and controlled the whole economy, selling to native blacks at extortionate prices, just to keep them in poverty. She remembers. And the others crawling on that beach remember too.

Suddenly, all eyes are on Nazish, and the survivors start crawling towards him. Some - well, those who still have feet - get up and walk to him, unsteady and exhausted, but determined. Before Nazish knows what is going on, they are on him, holding him down. He thinks they want to hug him as a thank you for being such a good TV presenter. He is wrong.

What happens next is captured in full colour high definition digital detail by Wellington - who they will not attack because he is native Numbian and, more importantly, he is holding the camera. The survivors want what they are doing to be shown on TV all over the world - they want their moment of fame. Because each and every one of them believes exactly what those watching them on television screens all over the world also believe - that if something is not on TV, then it does not really exist and so will not be real at all. They will make their own reality. And so they beat Nazish Naidu to death.

In the Presidential Palace, His Excellency Nelson Nasser Kenyatta Hitler Gerry Adams Bob Marley Prince Michael Jackson Blessed Bombobo, Supreme Commander and Protector of the Glorious and Noble Democratic Republic of the Ancient African Homeland of Numbia, is clapping and cheering at the images on his TV screen, a large glass of single malt whisky in one hand and a Havana cigar in the other. He laughs his huge laugh (one he has based on his hero Idi Amin) and lets off an equally enormous fart, as if in celebration at the turn of events and his country being on the world stage at last. Nobody can say that they're never heard of Numbia now!

In London, Alicia and Calvin watch what is happening

to Nazish Naidu on the giant screens at the outdoor event in horror. There are gasps from the crowd, and a few screams, but then someone cheers, and then another, and before long everyone is cheering, egging on the swimmers with every blow they land on Nazish's body.

Rasmus and Thursday watch the unexpected events on the TV in the mansion in Holland Park. Plans are already being made on how to manage the fall-out.

Anita, Debs and Sebastian watch at the X-TV offices, with other staff who are working late. They are shocked and saddened at their colleague's demise. But they have to admit too that what they are seeing is, despite everything, *great telly*.

Minty watches too. She has a look of stunned horror on her face, but not because people are dying for the sake of a TV show - which was expected. No, she is horrified simply because she is not allowed to make shows like this at the bloody BBC and knows - just knows - that this show will kill everything else in the ratings.

Oliver, Hugo, Lucinda, Sangeeta and Benjamin Cohen-Lewis are watching X-TV too, in their respective homes, sensing opportunity in Nazish Naidu's death. Many Members of Parliament watch too, merely for research purposes, of course.

Peter Baztanza is pacing up and down in his office watching X-TV but also keeping an eye on the other TV channels, just in case.

'Fantastic!' he says as the sharks rip the swimmers apart. 'I'm loving it!'

All over the world, from India to Indianapolis, from Brussels to Beijing, from Darwin to Dallas to Damascus to Dagenham, people are watching 'Shark Attack - Live' on X-TV.

This show is what they will talk about the next day at work, and it is what children will play in the playground tomorrow too - shark attack, just like on TV.

As Rasmus always says:

Art may imitate life, but life imitates TV.

Nazish Naidu takes a long time to die. The woman who is first on the beach - Charity, the winner of the race - finally puts the bloodied groaning and gurgling browny-red body on the beach out of its misery by smashing its skull in with a rock. She cannot bear to see such suffering, despite everything, and would do the same for anyone in the same situation. But she does not manage to do this before Nazish's eyes have been gouged out, his genitals torn off, and his tongue ripped out by the other swimmers. Wellington then watches as his fellow Numbians start eating the body parts - though out of revenge, hunger, ritual or all three, he is not sure.

Soon, the boats arrive, the helicopters land and the soldiers fire into the air to stop Nazish's body from being totally dismembered. Wellington stops filming and gets into one of the helicopters, as instructed. He has done his job to the best of his ability, and he knows that some of the shots he has captured with the hand-held camera are amongst the best he has ever taken.

Maybe he'll win an award for this? Maybe he'll become rich and go and live in The West? Maybe even America? Yes, he'll go to Hollywood, that's what he'll do, and become a famous documentary maker - maybe even win an Oscar one day!

This is the thought he is thinking when the soldiers suddenly grab him from behind and hurl him out of the helicopter into the shark-infested blood-bubbling sea below, just as they have been ordered to do by President Bombobo personally. Wellington knows far too much, so cannot be allowed to survive. It is better for his family for him to die this way - it can be called an accident and so President Bombobo will show his well-known great generosity by not killing Wellington's wife and children. The President knows Wellington would thank him if he could; but very different thoughts are in his brain as it bursts through the eye-sockets of his skull into the throat of a great white shark like sticky pink porridge.

Thursday turns off the television.

'The best so far,' says Rasmus, and Thursday grins. There is no way the BBC can compete with this.

*

PRESS RELEASE FROM X-TV

It is with great sadness that we announce the death of Nazish Naidu, X-TV's award-winning television presenter, who has passed away whilst filming in the African Republic of Numbia.

Arrangements have been made, with the generous help of President Bombobo, to return Nazish's body to the UK as soon as possible.

Nazish was a popular, talented and creative young man whose loss will be sorely felt by the entire television industry. It is, perhaps, some comfort to all Nazish's many friends and family to think that he passed away while doing the thing he really loved, presenting a TV show watched by hundreds of millions of people all over the world.

Our thoughts go out to Nazish's family – and his many friends – at what is a very difficult time.

We would appeal to the media to show respect for Nazish Naidu's family's privacy whilst arrangements for the funeral are made.

X-TV

*

Charity

Rasmus reads the press release and signs it swiftly and surely with his simple, elegant signature.

He knows that the substantial sum of money that is to be paid to Nazish's family, with the condition they make no further claim, will get them on-side. He also knows that because Nazish was killed in Numbia, no-one in the UK, and certainly not X-TV – which is broadcast from a foreign independent territory – can be held responsible for his death: that is the sole responsibility of the individuals who actually carried out the killing.

The authorities in Numbia are flying Nazish's body back to the UK today. X-TV has been warned that it is in not recognisable. In fact, it is in several pieces, with some parts missing.

Gary, Mercy and Danny Mambo are already on a plane which lands at Heathrow that evening. The survivors – including the winner, Charity – remain in Numbia and have been taken to a hospital in a convoy of rickety ambulances. They are very fortunate because the hospital they are being taken to is usually reserved specifically for the President and the ruling elite: it is the only one in the country which has a reliable electricity supply, sufficient stocks of medicine, and properly-trained doctors and nurses. When – or if – these survivors are well enough, they have been promised that they will be given a large sum of money, as well as a newly-built house in Numbia, and permission to leave the country if they so wish – and this has been promised by the President himself.

However, President Bombobo is an inveterate liar, and the fate awaiting the survivors is far less savoury. All of them, apart from Charity, will be patched up and bandaged in the hospital before being told they are being flown to South Africa for further expert treatment, and where they will be awarded a share of the prize money that all survivors have now been promised – something which makes many of them cry with happiness and thank God for their good fortune. In reality, they will be flown over the sea and summarily pushed out of the military aeroplane over shark-infested waters – to drown, if they are lucky.

President Bombobo will, of course, keep for himself the money paid to him by X-TV to be passed on to the winners

– which was his intention from the start. He has his eye on property in London and Paris, as well as some huge diamonds coming up for auction in New York soon, and will stash the rest in Switzerland with the billions he has pillaged from Numbia's economy already.

As for the winner of 'Shark Attack – Live' – the stick-thin girl called Charity who has two small children to feed and raise. When she realises she has won – that she is still alive but is now rich enough to build a future for her children by giving them an education – she just cries and cries and cries. She has struggled so hard for them, worked so hard too. She is happier than she has ever been, and can't stop praying and thanking God for his blessings. This, she knows, is her reward for keeping her faith, for going to church, and for giving some of her miniscule income to the ministry too.

She has been blessed by the Holy of Holies! Thank you Lord God Almighty! Oh thank you sweet Jesus! She is singing hymns in the hospital ward when they come for her, and she happily leaves with the soldiers who will take her to her new life.

But President Bombobo has a very special surprise in store for this child of Numbia – (and, as everyone knows, all Numbians are his children, whom he will always love and protect as father of the nation). Because this woman, who won the shark race, has brought shame on her noble homeland by killing the TV presenter on the island beach, smashing his skull in with a rock. And she also claimed in a TV interview before the race even began that she actually lived in 'poverty'.

This is impossible. Everyone knows that there is no poverty in Numbia. This is a fact that President Bombobo himself has declared many times, and which is proven by a completely independent government report showing the Numbian level of poverty at precisely 0%.

There is no problem with AIDS or illness in Numbia either, as official statistics also prove, and the country is so healthy that there are no disabled people at all – (though this is perhaps because they have all been fed to the sharks and crocodiles).

The great President Bombobo himself wrote the statute which made poverty illegal in Numbia - the first country in the world where poverty has been eradicated completely - and this is something which all children of Numbia learn at school and which makes them very proud, even the ones who do not survive to adulthood. The average Numbian life expectancy is thirty-three.

As for Charity…Well, she will be taken away by the soldiers to one of the President's palaces where, that very night, she will be gang-raped by the officers in Numbia's army until she dies, bloody and broken and whispering prayers to a God who has forsaken her. Her body will then be ground up and fed to the President's own guard dogs - which is an honour indeed.

'What would they do without a mother?' says the President with seeming concern at a dinner given the next evening to thank his government ministers for their loyalty.

On the table are what look like roasted piglets. But these roasted joints are, in sad reality, the succulent and crispy corpses of Charity's two young children.

President Bombobo thanks the good Lord Almighty for blessing him with the wisdom to lead and build the great country of Numbia to even more greatness. And, with that prayer still fresh on his lips, he sinks his teeth into a hunk of succulent tender flesh which used to be part of a two year old girl called Grace.

Animal Crackers

'Just look at all these brilliant animals,' says popular presenter Fiona Fudge to camera through her trademark smirk.

Minty, watching at home with Toby, groans inwardly. She has always hated Fiona Fudge, one of the new breed of cocky female BBC TV presenters whose ambition and inanity far outstrip their talent.

Animal Crackers is the BBC1 primetime show designed to go head-to-head with X-TV and grab back some lost viewers to boost the ratings. But nothing can hide the fact that it's all just a glorified pet show – and Minty knows just how dangerous it is to work with animals on TV, especially in pet-loving Britain, where the RSPCA was founded six decades before the NSPCC in the 19th century. Show a person being tortured and murdered in a TV drama, and you may get a handful of complaints. Show an animal being mistreated, even one unharmed in filming, even a prosthetic model, and the switchboards light up, producers are called into 6th floor offices, and apologies are issued as quickly as careers are ended. Minty can only assume that because the BBC has so assiduously avoided such animal shows for years, the DG has never had any personal experience of the vitriolic criticism they always generate. And yet the show managed to get the green-lights from the vast and labyrinthine BBC managerialist machine whilst experienced producers such as Minty were ignored. How? Why? Minty groans into her coffee.

The TV screen is alive with animals – cats and dogs and rabbits and mice and more – each in its own cage, miaowing and barking and sniffling and squeaking its part in an unnatural noisy animal chorus in protest at the heat and the noise and the new unfamiliar smells.

Fiona strides confidently amongst the participants in her long leather boots, just as she's rehearsed earlier. She knows that in an age of presenter-led TV this show is as much, or more, about her as it is about animals.

Establishing shots capture various angles of the pet show which are being spliced with the hand-held camera interview shots by the director, Neena Nana-Noonoo, and other production staff in the control booth high above the hall, to create what everyone hopes will be exciting live TV – but which is already starting to look amateurish in the extreme.

'Brilliant!' says Fiona, as she always does about everything, always.

Then Minty sees Hugo Seymour-Smiles on the TV screen, in the background, and she knows why he is there. Hugo loves the buzz of live TV being made – the whole adrenalin-drenched drama of it all – the excitement, the fear, the very smell of it! All TV people do, in a masochistic sort of a way. Minty knows how young and vital live TV can make you feel. As if it all mattered somehow...

Minty notices that there are bandages on Hugo's arms and legs – presumably covering wasp stings – and his mouth and lips are swollen to about three times their usual size. Suddenly, in the background, a security guard escorts Hugo off screen.

And then, for some reason best known to the director, the camera pans to show a young couple – white, with dreadlocks – hugging each other.

'Get those fucking crusty hippies off my television!' Minty yells at the screen. Toby ignores her – he knows his boss wouldn't want to miss a minute of this, so doesn't change the channel.

Minty wouldn't be surprised to learn that the two crusty hippies she is sneering at go by the names Moon and Starchild (formerly Keith and Louise) or that they consider animal abuse as bad at – or actually, like, mega worse than – anything that happened at Auschwitz.

On the TV screen, in close-up HD, Starchild hugs Moon, and Moon kisses Starchild, dreadlocks entangling with dreadlocks in a knot of crinigerous young love. Minty almost gags.

'Oh bring back Animal fucking Magic, all is forgiven,' mumbles Minty, her hand creeping over her eyes like a big snide spider.

The Funeral

The funeral of Nazish Naidu takes place at a vast municipal cemetery outside London. Rasmus is absent.

As luck would have it, Nazish made a will (as all X-TV staff are strongly encouraged to do) a fortnight before his death, little realising that his final wishes would be granted so soon.

Thus, Nazish Naidu, a British Asian from Wolverhampton and former Indian take-away delivery driver, enjoys a funeral which follows the traditions of New Orleans. According to his family, Nazish had probably got the idea because the day before he wrote his will he'd watched 'Live and Let Die' on the telly: it had always been his favourite Bond because it had so many black characters in it – or 'bruvvas', as Nazish always called 'his people'.

And so it was that the wailing jazz-infused strains of solemn and slow New Orleans funeral jazz rose melancholy into the overcast sky above the cemetery that day, as two hundred mourners trailed behind the shiny-black Rolls-Royce hearse in a cosmopolitan crocodile of mourning.

Colourfully-dressed Asian relatives from both Wolverhampton and Pakistan; jewellery-jangling rappers, mostly black but a few white and Asian, from the US and the UK; TV presenters aplenty, including Danny Mambo, Calvin Snow and Alicia McVicar, as well as several BBC presenters; various other celebrities from the worlds of reality TV, pop music and sport; and a representative of President Bombobo of Numbia – the head of the military, in full dress uniform, with a large white-plumed hat and a huge-bottomed wife who sports an even bigger feathery head-piece and colourful robes, and who is wearing some of the most enormous diamonds anybody there has ever seen.

Suddenly, the grim-faced man in a top hat plumed with two chicken feathers, who has been walking rhythmically and step-by-slow-step slowly in front of the funeral cortege, stops dead. He turns and his face splits into a grin aimed the jazz band behind him. They immediately cease their mournful plaintive dirge. Instead, and in a change as sudden as sunburst from behind a cloud, the loud and joyful strains of New Orleans jazz are trumpeted into the sky. This is a sign for everyone in the procession to start laughing and dancing as they follow the body to its final destination.

The music continues as the hearse draws up. Nazish's brothers and cousins lift out and carry the coffin - draped in a black power flag with a photo of Malcolm X in the middle - towards the cemetery chapel. Nazish had specified that he wanted a church service because, he said, he liked that 'churchy' sense of theatre, and the 'funky psychedelic stained glass windows'. And so, despite his non-Christian heritage, his funeral service was being held in the chapel. None of the family wanted this, and also would have preferred a more respectful low-key funeral, but they are all prepared to follow Nazish's wishes, especially as a celebrity magazine is covering the service and paying his proud parents well over half a million quid for exclusive access and photographic rights.

In the chapel, the celebrities do what celebrities are good at. They act the part, looking mournful and glum, yet with a wistful 'oh-what-might-have-been' glossy moistness in their eyes which, they know, will look just great in the photos when they're published later in that celebrity magazine. They know full well that going to funerals and memorial services for other celebrities is something that's great for boosting any celebrity's career profile.

From the walls of the church, statues and paintings of the Christian saints peer down dubiously at the saints of the present age: the celebrities from the world of television who glow holy and blessed right there in the half-light of the chapel, just as they shine brightly into the dullness of reality through our television screens, icons of a new age worshipped as much as any ancient deities or saints - but no doubt paid considerably more.

The service is taken by a bewildered-looking elderly priest, who somewhat unenthusiastically reads the memorial service text he has read a thousand times before. Then one of Nazish's brothers pays tribute:

'He…I mean…Nazish…my bruvva…was gonna be like well famous in America, and den dis ting happen, which is like *so* not fair and stuff, but as Nazish himself say, shit happen... sorry...bad ting happen...and then you die, innit?'

This seems a good place to stop, so he leaves the stage.

Suddenly, the solid phat sub-bass of a rap version of John Lennon's 'Imagine' booms into the room. Two men swagger forward towards the coffin - hardcore Eastside rapper '20 Cents' and his British counterpart '50p' who proceed to rap a version of 'Imagine' which features the word 'motherfucker' in every line and seems to describe the violent gang rape of a 'ho', followed by a 'gangbang' whereby two police officers are shot dead. Several members of the audience who can actually understand what is being rapped wonder if the original message of the song may have been somewhat lost in translation into its new urban idiom.

After this, the coffin is carried out to the graveside, followed by the mourners, but not accompanied by New Orleans jazz this time. Instead, a sound system blasts the theme tune for 'Live and Let Die' into the cemetery, followed by the Bond theme tune.

The celebrities look fabulous, dressed in their hats and finery. They have followed the advice of their PR agents in order to judge their dress just right - and some are even managing to cry wonderfully expressively, especially the ones who hardly knew Nazish at all.

Nazish's mother and father weep as they watch the coffin enter the earth, though no-one is sure if it is because they have lost their son in such tragic circumstances or because the rapper's version of 'Imagine' in the chapel was so ear-splittingly, tunelessly awful. Still, as more than one mourner ponders later, it could have been worse - it could have been 'Angels'.

One of the great things about the modern world is that if you die horribly on TV, you will not have died in vain, because you will have entertained people. And you will not be forgotten either, because the footage of your demise will be archived online and watched again perhaps millions of times in the future. You will, indeed, have achieved true immortality through your fatal fame.

As Nazish himself would have known, a death on television is never truly pointless. How can it be, when it makes you famous forever?

Because of the manner of his death, Nazish Naidu is now – at least for a short while – one of the most famous people on the planet, his fake Afro hairstyle copied by fans all over the world. It's what he would have wanted, for sure – he would have absolutely loved it!

Indeed, if he were still alive, Nazish would be well chuffed he was dead simply because of the fame his death had given him at last. He was now nothing less than an immortal, destined to exist for all eternity by the great God of television, granted everlasting digital life and fame.

Praise be!

*

Death. Most people fear it, yet there is nothing to fear. Why should there be?

After all, if we were not afraid in the vast eternity of time before we were born, then why do we need to be afraid of the world existing after we die?

I did not fear death when I stood on the roof of that office block, preparing myself to jump. Then it happened – the lightning hit me – and I was changed. I became Rasmus.

There is nothing to fear from change. Life changes. TV changes. Everything changes in time.

There is no reason to fear the future, because the future is now.

I am Rasmus, and this is my reality.

*

Nazish RIP

Today is the first meeting of all X-TV producers since the sad, bloody, squishy demise of Nazish Naidu.

'Welcome, people,' says Rasmus.

Everyone is there: Ravi, Anita, Debs, Sebastian - and Mercy and Gary back from Africa.

They have a lot to discuss today, but first, Rasmus knows he must address the issue of their colleague's death:

'I know I speak for all of us when I say that Nazish was a talented member of the X-TV team and will be deeply missed by us all,' he says.

Everyone nods sadly.

It might have been expected that some would blame X-TV for what happened, but two of Nazish's brothers have given TV interviews in which they fully support X-TV, do not blame the company at all for his death, and even suggest that this way of dying was what their brother would have wanted.

'I know that Nazish would have been delighted with the ratings for 'Shark Attack - Live', says Rasmus, turning everyone's thoughts into words.

'I flippin' well hope so, after all that malarkey, innit. We coulda been killed. I ain't never going back to that dump of a khazi,' says Gary Wu.

'Africa is a beautiful continent, yah,' says Mercy.

'Yeah, right,' says Gary with a shudder.

'He gave his life to be famous,' sighs Sebastian, perhaps sadly. 'How romantic! Like Rimbaud.'

'But Rambo ain't dead,' says Gary Wu, baffled. 'Is he?'

Sebastian ignores his Cockney colleague's ignorance with a queeny cringe.

'I'm sure it's what he would've wanted really, yeah,' says Debbie.

'Oh totally,' says Anita.

'Nazish was ripped to pieces by cannibals!' says Ravi,

unable to help himself from scoffing at the absurdity of all these eulogies.

'But they didn't totally eat him, did they?' asks Anita, unsure whether they did or not, 'because that would be like totally gross. Ewww!'

'What's it matter,' says Gary Wu, 'if the geezer's dead?'

'Twas TV that killed him,' sighs Sebastian, a wistful and distant look in his eyes. 'Twas beauty that killed the beast...'

Rasmus speaks:

'I am sure we are all very sorry indeed, and upset, about what has happened.'

Pause. Everyone nods solemnly at the floor.

'But we have a TV company to run. X-TV productions are underway – and we need to focus on making them as good as they can be, and increasing ratings. That is, after all, why we are here, and that is what Nazish would have wanted.'

With that, Thursday turns on the projector and shows the ratings as compared with the BBC and others. It is clear that X-TV has pummelled the competition again. It is way ahead of BBC1, with all the other digital channels lagging far behind.

Rasmus goes through the figures, before explaining his vision, detailing where X-TV is going and how it will get there.

When he sits down, the producers burst into spontaneous applause. Thursday grins and turns off the projector.

Nazish Naidu is gone and forgotten already.

In the past, it was always war that turned young men into memories. Now that is the task of TV.

Minty Mopes

Minty is sitting in her office looking through the 'overnights' – the ratings data showing the nation's viewing habits for the previous evening.

She has known bad 'overnights' before, but these are

disastrous – utterly appalling – and acutely embarrassing for the BBC.

'Fucking cocksucking wankers!' she shouts at the ceiling, despite knowing the words cannot possibly travel through the plaster and the MDF and the ceiling tiles to where the Director General is sitting in his office having just looked through the self same depressing stats. His head rests in his hands, eyes clenched shut, hoping reality will be different when they are opened. But the world is the same when his eyelids rise: it is the same dark and dismal place, with dreadful ratings for the BBC, and hardly any hope on the horizon.

It all comes down to 'Animal Crackers' now. If that succeeds, so will the BBC. If not, the Director General knows he will be expected to resign.

Two floors below, Minty too has her head in her hands – and, for the briefest of moments, she and Ben Cohen-Lewis are mirror images of each other, and would, if placed side by side by some enormous sky-descending hand, resemble a pair of gigantic bookends, hewn into the futile familiar shape of human despair.

But Minty does not 'do' despair – not really. She doesn't go in for 'if onlys' either. If only this, if only that – it's all just excuses by the weak, stupid and ineffectual. You get the hand you're dealt and you make the best of it. You don't moan and whinge and wish for a miracle with silly pleading prayers, or regret things in the past that you can't change. Why bother? It's just a pathetic waste of time. You get on with it and make sure you win through in the end, through hard work, determination and ruthless ambition. *Life is real, life is earnest.* Such is the philosophy of Minty Chisum, Head of Entertainment at the BBC, who got where she is the hard way and on her own merit – which is more than most in the broadcasting business, as she well knows.

However, at this moment – a moment of failure and sadness and the approaching doom of a ridiculous live pet show – Minty cannot help herself from succumbing to the heavy weight of depression at her present reality. And so, just for a moment, she even allows herself a couple of 'if onlys' too.

If only they'd agreed to her plan to get Peter Baztanza on board. If only they had done it her way from the start. If only she were in charge of the whole, huge, messy, corporate, bureaucratic muddle called the BBC, then things could be different – and they *would* be different too. Oh yes, they'd be different alright. The programmes she could make with Peter Bazataza! The ratings they would get! Then they'd be really competing with X-TV and Rasmus Karn, that upstart from nowhere who has, inexplicably, managed to transform himself into a media mogul, start up a TV company and hammer the BBC in the ratings in a matter of a couple of months.

How the hell did he do that? How?

'How the *fuck* did that happen!?'

Minty suddenly realises that she has been talking to herself, not thinking her thoughts quietly at all. The first sign of madness, they say – though *they,* as usual, are wrong. The first sign of madness is wanting to work in TV, or just thinking that fame and celebrity and television matter more than real life.

Minty really hopes that no-one is going to bother her this evening about 'Animal Crackers'. If they do, she'll refer everything back to Hugo or Oliver or Benjamin Cohen-Lewis himself. As far as she is concerned, these idiot men – this pale, male, stale brigade – came up with this crap, so they should take the rap for it too. It is Minty's intention to wash her hands of the whole sad business – as much as a Head of Entertainment with overall responsibility for the show can, that is.

Hammersmith

At the X-TV building in Hammersmith, Rasmus knows Animal Crackers will be a complete and utter ratings disaster. Of course it will, and it is obvious why too.

A pet show of any type is not something X-TV would ever

go near, and not just because the ratings would be miserable. With the exception of the X-TV sex shows using animals (which the four-legged creatures in question do seem to rather enjoy) and the programmes showing animals attacking humans (such as 'Shark Attack Live'), Rasmus would never ever approve of any pet-related show being broadcast on X-TV. The concept could work in some parts of the world, and Rasmus has not ruled it out for those markets. Perhaps the Chinese, for example, would even actually quite like to see live animals - of all types, including snakes, turtles, sharks, as well as cats, rats, bats and dogs - being freshly killed and prepared by an expert chef and turned into interesting delicacies. But the British? Never!

Rasmus knows that the one sure-fire way of alienating, upsetting and offending the British public is through hurting animals, and any UK TV show that has ever done this - or even pretended to do this with no animals actually hurt in reality - has received more complaints than any others. This is perhaps why BBC management think that 'Animal Crackers' has such taboo-breaking ratings potential.

Animals are different - and animals which are pets are, in televisual terms, untouchable. Any British TV channel that ever dares to hurt a pet, especially a cat or dog, is writing its own suicide note.

Rasmus announces a new bonus for all producers and an increase in investment. Several new shows are in production, some follow-ons to previous shows, such as the various 'Celebrity Suck-Off Specials', which Sebastian has made his own with top-quality and original production ideas, and various other sex shows involving grannies, dwarfs, amputees and as many combinations that the producers can think of. X-TV also hopes to revive the game show format - a new show which will give contestants the chance to win huge cash prizes, but not before they endure qualification rounds which will test even the most determined attention-seeking TV *sleb*.

The meeting ends with the producers delighted at the way things are going. Mercy and Gary are given immediate leave, and a free holiday break in the South Pacific, to thank them for their hard work on 'Shark Attack – Live' and to reward them for the trauma of all that happened in Numbia.

Ravi follows them out. Rasmus has noticed how quiet he is, but knows too that there are bound to be moral conflicts in the minds of the X-TV producers. After all, if he were the same man he was before, he would be conflicted too. But people accept everything in time.

Anita and Debbie leave together. They are a good pairing – Rasmus has noted good chemistry between the two.

Sebastian is alone, as always – and, as always too, seems to be recovering from the previous night's binge of drugs and/or booze and/or sex – and probably all three. He is a loner, always separate and aloof, but Rasmus knows that this is no disadvantage – in fact, quite the opposite. It is a distinct plus, creatively speaking. Loners are always the most creative people on any team.

'Could I purleeeeease work with that beautiful boy, Calvin,' he says, looking lovelorn, 'preferably as intimately as possible...'

'We'll see what we can do,' says Rasmus.

The X-TV building – that fourteen story block in Hammersmith – was empty when X-TV first moved in, initially on the top floor. Its former occupant was a large corporate that had gone bust in the 2008 crash. As X-TV expanded, its staff expanded too, so that as well as occupying the fourteenth floor, it also came to occupy the thirteenth floor, then the twelfth, and then the rest.

Rasmus looks out at the city from his high window. Thursday stands by his side. It is a hot day in May, though pleasantly cool in the perfectly air-conditioned office, and a mist seems to rise from London like a huge new ghost made of all the better-forgotten memories of the city's past.

'Soon,' says Rasmus, as Thursday looks out at this strange

foreign city with its strange dangerous tribes and even stranger traditions.

Not long now.

*

What is fame? Something valuable and desirable? A guarantee of immortality?

Or merely the advantage of being known by people about whom you know nothing and care even less?

Whatever the truth, fame is desired – yearned for by millions, by billions even, in every country and culture.

Perhaps fame is so precious because it is so intangible, so invisible, summoned into existence with sorcery and magic tricks.

Certainly, oblivion is the rule and fame the exception of humanity.

And what of celebrity? Is that the same as fame? Or its deformed runt of a brother, perhaps?

TV created fame and celebrity in their modern guises. All those who crave fame know this truth. They all know that TV is reality: it is our lives that are unreal.

And with TV comes fame and life after death.

Eternal life: that is the bright burning flame of fame.

I am Rasmus, and this is my reality.

*

The End of the Beginning

London looks ill. A lazy haze of heat hangs weary in the tired air, stifling the streets and pavements, sucking the breath from its inhabitants, and hissing out a vast polluted cloud of mist that makes the whole city look a sickly shade of yellow, like the belly of a dead fish.

It is early evening. Commuters are sweating on the tube

and the buses, weary at the weight of the weather which seems to be pushing them into the ground with a force greater than gravity. They read newspapers and magazines, listen to music on their iPods, or just stare blankly at the dull world around them. Some of them speak their monologued inanities into mobile phones. All these people really want is *not* to be where they are, but to be in their homes - big or small, rented or mortgaged, elegant or shabby - where they can get away from the rest of humanity and do what they've been waiting to do all day: turn on the TV.

They will stare at their screens - on TV or laptop or notepad - as they always do, transfixed and entranced, as if watching bubbles forming in the primaeval ooze, not thinking, not doing, not being. Just watching TV. And those who don't live alone will prove the maxim that, if nothing else, the invention of television has proved that people will look at anything rather than each other.

Inside the Exhibition Hall at Olympia, where 'Animal Crackers' is filming, the smell is more than over-powering. It seems solid in its force, a shockwave in the air. The heat has been cooking the hall all day long, and with it all the people and animals in it, with their sweat and stench. Now, at the end of the day, as the pets are about to be paraded for the viewing public and the judging to take place, another smell has been added to the mix - vomit.

It is the smell that does it - the iodine stench of animal shit and sick everywhere in the hot air - and families who had gone to the Animal Crackers pet show for a nice day out, not to mention a chance to get on the telly, are the first to be affected by it, with children puking up at regular intervals onto the floor. Throughout, the cleaners do their best to mop it up, but the smell lingers in the air like an acrid poisonous gas.

Fiona Fudge watches all this with an appalled - yet smirkily smug - look on her face. There is of course no way on earth that a BBC presenter like her, (on a large six-figure package), or any other BBC staff come to that, would ever actually help to clean the mess up themselves. So the security guard and others are asked by the BBC production team to help in mopping up the puke-splattered mess, but they refuse. They are then told that if they do not help then that will be considered insubordination and they will be sacked. So they get busy with the buckets. And now, a stench of disinfectant is added to the shit-piss-animal-vomit smell of the hall.

The Somali security guard curses the vile and disgusting infidels around him, and prays to Allah that they will all burn in agony for eternity with Shaitan in the unholy firey furnaces of hell.

Hugo watches all this with a growing sense of doom, like a passenger on a huge beautiful ship sailing inexorably towards the inevitable, merciless iceberg. It is so hot that he can hardly breathe – though this is not exactly helped by his blocked and swollen wasp-stung airways.

Fiona Fudge is ready in front of camera, smirking away and trying to stop her nostrils twitching at the vile stench all around her. 'Five, four, three' says the production assistant, mouthing the two and the one, which is Fiona's cue to take a long deep breath. The assistant points at the presenter when the camera goes live. And then, they are being beamed live to the nation on peak-time BBC1.

'Welcome to Animal Crackers, coming to you live from Olympia.'

Cue the inanely unoriginal and annoying plonkety-plonk theme tune which, predictably, features sampled animal noises – barks and miaows, but also baas, moos and even cock-a-doodle-doos.

'We'll be showing you some brilliant animals this evening – let's go and meet some of them now, shall we?'

Fiona Fudge strides over to the finalists and their pets at the precise moment that the two crusty hippies who so affronted Minty decide to initiate their direct action to liberate the poor oppressed animals from their fascist human abusers.

'I love you, man,' says Starchild.

'I love you too, babe, kindathing,' says Moon.

'Stay chilled yeah,' says Starchild, 'at peace with the universe.'

'Yeah, peace, kindathing,' says Moon.

'Brilliant!' says Fiona Fudge to a hamster.

And then it happens.

With a noise resembling the rattling of gunfire, bundles of lighted firecrackers are thrown across the room, together with stink bombs that start emitting foul-smelling and colourful smoke into the air with a deafening noise.

Everywhere, people are running this way and that, screaming and crying in panic. This makes it much easier for Starchild and Moon to get on and fulfil their mission. Cages are opened. The incarcerated are liberated. And animals do what animals always do when confronted by noise and fire and chaos – they panic and attack.

The director can see what is happening and barks into Fiona's earpiece:

'Get out of there get out of there!'

But she also makes the decision not to stop broadcasting, hoping – beyond hope – that something can be salvaged from this disaster.

Fiona, meanwhile, is wandering round in a daze, disorientated by the noise and smoke, as indeed is Hugo and everybody else there. This means they all bump into each other, collisions which on occasion result in serious bloody injury.

The Somali security guard has had enough. Pushing everyone out of his way, he makes for the exit, punching a fellow guard in the process. He emerges out into the fresh polluted air of West London with the billowing smoke from the hall. Hurling his cap skywards and ripping off his tie, he runs as fast as he can away from that place. He has decided that he'll be much better off – not to mention safer – if he returns to the failed state of Somalia than in this infidel hell of a crazy country.

Before long, the noise of the dying firecrackers is replaced by that of the animals, whose horrible and blood-chilling cacophony of barks, hisses, screams, growls and squeals is as ear-splittingly loud and primitive as any daytime TV talk show audience.

Beasts of all descriptions are running around attacking anything that catches their eye, startled by their new freedoms and spooked by the fireworks and noise. Dogs are chasing cats, cats are chasing mice, and they are all chasing rabbits, who are running for their lives as they dash and slalom between the people, tails bobbing white with every hop. And the newly liberated snakes are hissing and spitting as they slink quickly out of their tanks, their fangs sharp and primed.

The base animal survival instincts of the jungle kick in immediately. It is fight or flight - and possibly, both. But with most exits closed, and the ones that are open obscured by the smoke and chaos, the flight becomes fight almost immediately. And so the attacks begin on the panicking people.

Fiona is one of the first to be mauled. This is because she is one of the first to reach the main exit which, due to the proximity of their cages, is now being guarded by several dogs - massive and fierce Rottweilers, Alsatians, dangerous-looking bulldogs and pit bull terriers - a pack of wolves focused on the hunt. They are particularly attracted to Fiona's long leather boots, which look and smell very tasty to their hungry noses. As she approaches the exit, a Rottweiler bounds onto her and clamps her neck in his enormous jaws before she can even scream.

Before long, the shiny sausaged innards of Fiona Fudge are trailing out over the exhibition room floor, together with what seems like huge amounts of her bright red blood, especially after a dog bites through the artery in her neck. This makes the ground wet and slippery, causing more people heading for the exits to fall over and so become easy prey for the dogs.

Meanwhile, Hugo takes refuge below a big glass tank. It is, in retrospect, not one of his better ideas.

'Ow! Um! Ow!' he says as a tree-green snake sinks its

fangs into his already-swollen nose. Hugo crawls out, the snake dangling off his face, looking as if he is a newly created chimera stitched together by some mad Victorian biologist in a laboratory. If he had realised that the snake actually meant him no harm, was not venomous, and was perfectly happy to stay there all day affixed to the end of his nose, then he may well not have decided to emerge from under the tank, especially as the snake's body wriggling and writhing in the air makes it - and Hugo - irresistible prey to all animals who see it. Hindsight is a wondrous thing - thinks Hugo. Much later. In hindsight.

Before long, Hugo is on the floor, covered in a furry blanket of cats and dogs, each of which has either got its teeth or claws or both sunk deep into his flesh. Hugo, weighed down with pets, is unable to move and expects the worst - a feeling intensified by the repeated ramming insertion of a dog's penis into his mouth.

'No! Bad dogs! Come on, man, chill out, yeah? You're free at last, yeah!' yells Starchild, somehow believing that dogs and cats can instinctively understand English when spoken by dreadlocked animal rights activists from Surrey.

Within seconds of Starchild's pleas, the dogs and cats transfer their attention from Hugo to her, and then they attack Moon who has, apparently, come to the rescue.

'Chill out kindathing! Chillax!' says Moon to the animals. They will be his last words.

'Maybe we should have thought this through a bit more carefully, yeah?' thinks Starchild as a dog bites into her throat. It will be her last thought in this world, as the puncturing teeth of an Alsatian put an end to her dreams of inter-species harmony on an animal-loving earth.

Hugo manages to escape and find another cage, some distance away. He crawls under it. Suddenly, a Chihuahua leaps up to his face with a squeaky yap and attaches itself with a little growl to his bloody and swollen nose with the snake. Hugo yowls in pain but can do nothing, so he squeezes himself tightly into the space under the cage and thinks nice thoughts, holding the Chihuahua in place so it doesn't bite him anywhere else.

The noise of the animals barking and growling - and the people screaming and wailing - is soon joined by the noise of sirens: first, the fire alarm that someone eventually sets off, and then the shrill American-esque screeching of the police and ambulance sirens outside.

Riot, Not Riot

Outrage is felt throughout the land that the torture of animals is being broadcast on TV – and on the BBC too, at that – outrage which, in the swelteringly tense streets of West London, soon turns from simmering unrest to violence.

A fourteen-year-old boy called Malcolm throws the first brick that breaks the first window that starts the first eruption of violence. His single mum, who was fifteen when he was born, named him after Malcolm X and would be proud at his brick-shaped protest against this racist society - that is, if she wasn't lying fast asleep on a stained sofa in her council flat in White City, stoned out of her head on skunk, being roughly fingered by one of her many men-friends whose name she has never even asked.

Before long, the streets of West London around the BBC erupt in violence – though this will be classed as 'a disturbance' and not 'a riot', a neat trick by the government to avoid paying for the clean-up (the insurance companies will have to foot the bill). Police reinforcements come in from all over London to stop the 'non-riot' from spreading.

It is only when it is clear that the focus has shifted away from the Exhibition Hall to the streets that the decision is made to stop broadcasting the footage of 'Animal Crackers' and to end the show.

Fiona Fudge lies dead on the floor, her face a bloody and

gnarled mess of flesh from which her once-smirking lips have been bitten away. There are groans and crying from the injured, a great whimpering of man and beast, like a scene from a battlefield in some strange inter-species war.

Minty watches the unfolding events on TV in the living room of her home with a feeling of joyful expectation rising inside her. A gleeful grin tugs at the edges of her mouth, despite everything.

She knows that this is the end for the Benjamin Cohen-Lewis, and she knows that she would be perfect for stepping into his shoes. But patience, she thinks – patience. If she makes a move now, it will look as though she is too eager – too desperate – especially with bodies still warm on the floor, and the fate of so many unknown.

Besides, the Director General's resignation hasn't been announced yet – but it certainly will be, and soon too. Minty will wait until the announcement on the news. Then, with precision timing, she will strike.

But in his sixth floor office in BBC TV Centre, Benjamin Cohen-Lewis is not worried. Such is his utter failure in his present role that he has already been offered several lucrative teaching contracts at prestigious universities in both the UK and the US. He sits admiring himself in his hand mirror. Wouldn't look too bad in a beard, he thinks – intellectual, even. The soon to be ex-DG is keen to adopt the pognophiliac tendencies of the average middle-aged mediocre academic, so decides to stop shaving there and then. Well, you have to look the part – don't you?

Rasmus knows he is witnessing the implosion of the BBC. He knows too that Minty is out there, watching and waiting.

The 'non-riots', though initially confined to West London, worsen. Crazed mobs, as bold as they are lawless, trash everything in their way to get their hands on what they want.

Rasmus notices, with a wry smile, that the looters' first choice of items is a flatscreen television. Shot after shot on the live coverage shows looters lugging away these huge, expensive items, smiles of delirium and delight on their faces, their eyes wide as TV screens. No doubt they are looking forward to watching themselves on their newly-looted widescreens later. Rasmus knows it is a victory for television – the triumph of machine over people.

The 'BBC riots' is what they are calling the disturbances – it's a name that an ITV (and former BBC) newsreader came up with and which seems to have stuck – and the pictures are being beamed all over the world.

Rasmus goes onto the roof of the X-TV building and watches West London burn.

And so the people riot. But who can blame them for protesting against the constant betrayal of their hopes and dreams?

Really – who can blame them?

They want to be on TV, and so now they are, albeit with their faces half-covered. For many of these millennials, it's the best thing that's ever happened to them in their short, disappointing, pointless lives.

Out over West London, distant sirens scream and wail into the air, as if strange aquatic creatures in the river are calling a mate. A bitter orange glow rises into the night sky from the flames and streetlights. The smell of smoke and ash hangs in the air like a shroud. The city looks as if it's been bombed. Again.

Less than a mile away, Minty is standing outside her house in Hammersmith looking in the direction of the X-TV building and Rasmus himself. For the briefest of moments, Minty thinks she can feel Rasmus there, beside her. She knows he is out there, enjoying everything, breathing it all in.

Rasmus stands on the roof, in the place he stood on the day his life changed, tasting the heat of the night and listening to the muffled cries of sirens and looters rising from the rioting streets below.

The air is hot with the scent of fire, and it smells sweet to the nostrils of both Minty Chisum and Rasmus Karn.

It smells of the future.

It smells of change.

*

Human nature never changes – it is the same now as it was when people worshipped demons, stood in awe at the cruelty of kings and despots, cut out the still-beating hearts of their human sacrifices, garrotted and tortured each other for fun, and enjoyed eating each other's flesh. People are people, so behave as people always have.

There has been no change in human nature. Not really. The brain is the same, as are the instincts and desires. The only change is in the details, the performance, the reveal.

And so, the beginning is at an end, and soon, the end will begin.

Soon, X-TV will dominate the airwaves throughout the whole world: Europe, Asia, Africa and America. All these things will come to pass, in time.

But for now, in this city, on this night, it is enough to watch the world burn, to know that things will never be the same again, to see and hear and think a new future into existence.

I am Rasmus, and this is my reality.

PART TWO

THE MIDDLE

Rain

Rain. Hard rain. Lashing, drenching, torrential rain. Cold black rain for a cold black city. It is London in July and this is an English summer. It has been raining for days.

The sky is the colour of dead television.

Rasmus looks out at the landscape: the black muddy smudge of the Thames bubbling in the downpour, the dead-looking leaden clouds, the misty concrete blur of the horizon – and the rain ramming down like nails.

It is the weather that stopped the disturbances in the end. Not even hardened criminals or radical revolutionaries like getting wet and soggy – not when they can be at home watching themselves on their newly looted widescreen TVs.

But, despite the downpour, the air still fizzes with tension, the fuse of a firework seemingly extinguished but still hot and quick, slumbering silent under the surface, and ready, when the time comes, to explode.

Time, just time. That is all it will take. And Rasmus knows it.

Minty Chisum is now Director General of the BBC. She sits in the DG's office – *her* office – breathing in the power. Then there are the perks – the salary and pension, the chauffeur-driven car on call 24/7, the silverback status which makes everyone want to defer to her, to be sycophantic, to do *exactly* what she tells them. There is fame of sort too – headlines on the TV news, features in the nationals. They're even calling her a feminist icon and a role model for schoolgirls on Radio 4. But then, they call pretty much all high-profile and/or celebrity females feminist icons these days at the BBC.

Minty takes a deep nostril-filling sniff of the air in the office, to let her lungs suck it all in just a little more. She summons her secretary – one of three – with the press of a button.

A knock and the door opens.

'Kirsty, schedule a meeting with the usual team.'

'It's Carly, not Kirsty, Director General.'

'I thought Carly was doing the photocopying?'

'No, that's Kelly.'

'Not Kylie?'

'There is no Kylie,' says the secretary, who close-up looks older and uglier than Minty first thought, acne scars and age cracks now visible through the layered foundation face paint.

'Kirsty, Carly, Kelly, Kylie, Cunty – why should I give a tuppenny fuck what you're called?'

Silence. Carly had never heard a Director General use such language before, so she will pretend she's heard nothing – a common response at the BBC. Almost a tradition, in fact.

'I mean,' says Minty, 'why can't people have proper fucking names any more anyway – as in, names not lifted from some cunting daytime TV soap opera?'

Carly has never heard the word 'cunting' before either. She isn't really sure if it's a real word – she'll have to have a look on her Spell Check later. For some baffling reason, at that very moment, colourful images of bunting start floating through Carly's obedient mind. Bunting rhymes with cunting, of course, so that must be why. But then why didn't she think of hunting? Or punting?

These are the tiny insignificant secretarial thoughts now being copied and pasted into Carly's thought processes – so much so that she finds herself unable to utter a word.

'Carly, Crappy, Cunty, Colostomy, Cock-sloppy – I don't care! From now on, I shall just call you all 'secretary' or 'you' or maybe just 'cunt' – or just give you each a number – no need for names really, is there?'

Carly nods, leaves the office and schedules a team meeting as instructed.

Meet the new boss...

'Mucho congrats, Minty,' says Oliver Allcock, when he eventually arrives – late as always, bursting through the door like a fart. 'It really is an incredible achievement, and to think, this place used to be a lady-free zone a few years back.'

'It still fucking is,' says Minty with a grin.

She has never knowingly met a lady in all her time in broadcasting – plenty of utter bitches, shameless slags and cruel heartless cunts. But definitely no ladies.

Toby, who stands loyally by her seated figure, like one of Cleopatra's slaves, positively purrs in delight. He knew if he stuck by Minty, she'd see him right. And now he's personal assistant to the Director General of the BBC, a new post created especially for him! Minty's just like Bette Davies, he thinks, but younger, better, crueller, and not a smoker or American either, which is a definite plus.

'Yes, um, many congratulations, Minty,' says Hugo.

'You've already said that, um-Hugo,' says Minty, holding up a hand to Lucinda and Sangeeta, in case they want to repeat their gushing congratulations too. She doesn't know why exactly, and cares even less, but Minty has always enjoyed criticism more than praise.

'I, um, know, Minty, but the meeting hadn't, um, started when I...um, not until Oliver arrived.'

'If you are *ever* fucking late again for one of *my* meetings Ollie, you're out – d'you understand?

'But Minty...'

'Don't you *but Minty* me. Do what I say or you're fucking finished – got it, Allcock?'

'No problemo,' Oliver nods.

Toby smirks. Lucinda and Sangeeta peer warily at Oliver

Allcock, Controller of BBC1 – and edge their buttocks slightly away from him in their chairs. They suddenly realise he could now be the next out – not Hugo, after all – and certainly don't want to be too closely associated with a loser.

'Now then,' says Minty, 'I...'

'Minty,' says Hugo. Everyone stares at him. 'I wonder if, um, I might say something? It is, um, rather important…'

Minty sighs. She has known Hugo forever and has never known him say anything important, not for many years anyway. But it's easier just to let him get whatever bureaucratic bee he has in his bonnet out into the fresh air, where it can fly away unnoticed, then shrivel up and die useless and unloved in some sad dusty corner of his mind, than to star t a conversation – which, with Hugo's hesitant speech impediment, could literally take hours.

'I have, um, decided, um...to...um...resign.'

Everyone looks at Hugo. This is an unexpected – though not unwelcome – surprise to all present. But everyone also looks puzzled – probably because no-one can think of anything else Hugo could possibly do for a living. But also, because quitting now might jeopardise his retirement fund, and no BBC lifer would ever want to do that – which is why so many stay exactly where they are, for years and years and years…

'There,' says Hugo, 'I've said it. You see, um, I feel as though I have, um, come to the end of a long chapter in my, um, life…and I'm rather keen to, um, start a new one.'

Minty knows exactly what he is going to say.

'And, um, Felicity wants a divorce, so…'

The new DG smiles tightly – she has long expected this.

'And what with the, um, eye…'

The eye? Surely he means 'the eyes?' thinks Minty – the eyes of the BBC Head of Vision that can see bugger all.

'What about the eye, old boy?' asks Oliver.

'Oh, sorry, um…must have forgotten…I'm blind, um, in this eye now.'

Hugo points to his right eye socket. Everyone leans in to take a look. It looks almost identical to the other eye, slightly more bloodshot perhaps, and duller, but otherwise no different.

'You see, um, I had a little accident...'

Hugo proceeds to explain how he had been gardening at home, planting organic tomatoes in pots and tying the fragile plants to little green canes for support, when he bent down to plant another and one of the aforementioned little green canes stabbed him straight in his eyeball. He went to hospital, clutching a handkerchief over his eye, and was told that his sight could possibly be improved by a quick operation. Unfortunately for Hugo, it left him blind in that eye instead, though no-one told him exactly why that had happened, or that the surgeon operating on him was actually a student trying the operation for the first time...

Needless to say, the hospital which blinded him has won numerous awards for its unrelenting efficiency and high standard of patient care.

'And so, um, what with the blindness in one eye, and retirement looming anyway, I thought to myself, um, why not go now – and live the, um, life I want to live?'

Oh how Minty wished a whole lot of other BBC lifers would do the same. But no, they hung on, limpets on the rock, clinging on for dear life, rotting away uselessly, creating nothing but a great big putrid pong, as immovable as stone.

'I told Felicity, and, um...that's when she said she, um, wanted a divorce.'

Despite never having married, or ever wanting to, Minty knew that what really mattered in a marriage, and thus a divorce, was not friendship, companionship, or that temporary weakness – and perpetual mental illness – called 'love'. No, what mattered, always and in every single case, was money.

She had met Hugo's wife a few times – saw on each occasion the hunger for status and wealth in her eyes – and knew immediately that Felicity would grab all she could from Hugo and suck him dry. Minty had witnessed the change: Hugo's wife had been chipping away at his confidence for years, grinding him down to dust, until there was almost no man left at all.

After the usual commiserations and false bonhomie, Minty gets down to the business of discussing what she intends to do as DG and what changes she has decided to introduce. If the BBC is to compete with channels like X-TV, as well as all the others, online or digital, it needs to stand out. It needs, in (just two) other words:

'Peter Baztanza,' says Minty. 'He's in for a meeting tomorrow.'

'Incredible!' says Oliver Allcock.

'Cool! That's a really great idea, Minty,' says Lucinda, and Sangeeta agrees.

Hugo nods. He knows that this BBC – a BBC of Minty Chisum and Sir Peter Baztanza – is no place for him. Perhaps it never was. Perhaps he has lived the wrong life in the wrong place with the wrong wife and family too? It's all wrong, all right!

'I wish you, um, the best of luck, Minty,' says Hugo, bumping into the desk and knocking over a plastic pen container as he reaches to shake the new Director General's hand – something that almost makes Toby restrain him, like a bodyguard protecting a star from a fan.

'Thank you, Hugo. We'll miss you,' Minty lies.

Hugo stays behind with Minty for a while after the others leave, though he refuses the drink Minty offers – having one eye is disorientating enough without whisky-wooziness affecting his balance even more.

'Wales?' says Minty. 'What the fucking hell do you want to go and live in Wales for?'

'Well, um, you see, I've been offered a sort of job, um, teaching part-time at a local university in the South Wales valleys…and, um, anyway, I saw some really lovely properties there, um, with land and really, um, rather reasonably priced, so I thought, why not go and live in Wales, and, um, become an organic farmer, um, just for fun really.'

'Fun? *Fun?!* Oh for fuck's sake! Have you *any* idea what life is like in fucking Wales?'

'Well, um…'

'Wales is a miserable fucking rain-soaked place – especially the valleys, where the only growth industries are mass unemployment, rampant drunkenness and almost compulsory drug addiction, and where the only thing manufactured these days is complete and utter fucking unabated mind-rotting misery.'

'But, um…'

'And the Welsh! The Welsh are a race of swarthy phlegm-dribbling drunken drug-addled chapel-chasing fucktard dwarfs who live – and I use the word loosely – in a land of perpetual fucking rain, misery and failure, where the only fucking thing to look forward to is death itself.'

'But Minty,' Hugo protests, 'Wales is a beautiful country.'

Hugo loved Wales. He loves it too, and will always love it – the mystery, the magic, the cheap house prices. He loves it not only for the scenery but for the whole concept and idea of the place – the Celtic *cwtch* of Cymru. Indeed, he loves it to what, he knows, the famous 16th century Welsh mathematician – and creator of '=', the equals sign – Richard Recorde might have called a zenzizenzizenzic degree.

'Beautiful Welsh valleys?' scoffs Minty.

'Well, um, in their way, yes… and the property is so inexpensive too – the divorce settlement won't, um, leave me all that much really, what with, um, Felicity's needs, and the girls' school fees.'

Minty leans forward, like a curious psychiatrist studying a particularly pin-headed patient encaged in some Victorian asylum, speaking slowly and clearly so as to be understood by the unfortunate mental defective before her:

'The reason, Hugo, why it's so fucking cheap to buy a house in the Welsh valleys is because it's so fucking awful nobody wants to fucking live there!'

'Well, I do,' Hugo snaps back, um-less and free.

'Oh my fucking good God,' says Minty. 'You've already bought a place, haven't you?'

Hugo smiles and blinks his one good eye. He is at peace at last.

To think, after so many years at the BBC, mostly in this very building, he is soon going to be free - to be the man he always wanted to be, no longer worried about what anybody thinks, no longer trying to please everybody else - his father, or his wife, or his children - and no longer all that concerned about money either.

It was as though moving to Wales would herald his rebirth, his parthenogenesis, his liberation into his new wonderful Welsh life as an organic farmer and a lecturer in media studies (part-time) at an up-and-coming, vibrant and forward-looking university (at least, that is what it says in the glossy prospectus).

It is somewhat unfortunate that, after leaving BBC Television Centre that day, Hugo Seymour-Smiles walks straight into the path of a bicycle - (it's just so hard to judge distance with one eye!) - and is then punched in the face by the lycra-fied cyclist whose vocabulary range seems to start at 'fuck' and end at 'off'.

After a quick visit to the hospital, where a bandage is wrapped around Hugo's head and his bruises, cuts and newly darkening black eye are seen to, he heads off home to start packing up.

'Wales here we come!' he thinks to himself as he starts filling box after box, singing *Bread of Heaven* with half-remembered words to get him in the mood.

After a while, he puts on a CD of Shirley Bassey, singing and dancing along with the girl from Tiger Bay as she belts out *I am what I am, Goldfinger, Diamonds are Forever* and *Big Spender*, at such a volume that the next-door neighbours think that perhaps Hugo and Felicity have already moved without saying goodbye, and have been replaced by the kind of fashionable gay dinkie couple of the type that can afford the house prices in Twickenham.

'Free at last, um, free at last,' chants Hugo, deciding to read the letter from Felicity's solicitor in full later, 'great God almighty, free at, um, last...'

*

Everything changes; and the more things change the more they stay the same. Existence does not change. We live, we die. The End.

We are living through an age of transition, when one system will be replaced by another. It is nothing less than the evolution of our species.

Has democracy and the right to vote really empowered all individuals? More than television has, for example? More than the internet? More than X-TV?

There has been rioting on the streets; the economy is in freefall; useful work and hope are drying up. So why is it surprising that people seek succour by retreating into an artificial world of TV or the internet?

And so a technology invented to learn about the world has become the means through which we tell the world about ourselves, as it was always inevitably destined to do.

Sometimes, reality is too real to bear. Sometimes, the alternative reality on offer is far preferable. So why not choose pleasure over pain? You know you would. You know you do.

The first stage is complete. Now, the second is starting – one which will, after a great deal of suffering and sacrifice, create the future for us all.

The future we all want, the future we all need, the future we all deserve.

I am Rasmus, and this is my reality.

*

From: Hugo Seymour-Smiles
To: Minty Chisum; Oliver Allcock; Lucinda Lott-Owen; Sangeeta Sacranie-Patel
Subject: Retirement Do

Good afternoon, everyone.

Just a quick email to say how much I have very much enjoyed working with all of you, and that no doubt I shall miss you all when I am living the life of an organic gardener. Back to the earth and all that!

Anyway, I have already emptied my office, and the house has been sold too – for a knock-down price, but then it's a buyers' market these days, isn't it, what with the recession, and Felicity wants things to be over with quickly – so I am no longer in London. All very sudden, but it's best to move on, isn't it? Always best to avoid going backwards, going forward.

I am actually writing this email on my laptop whilst parked near the Severn Bridge en route to South Wales. Just waiting here for the RAC, who will drive me to my own car on the M4 hard shoulder, and fill it up with enough petrol to reach a service station.

Silly, really, but I simply failed to see that the petrol was running out on the dashboard – too many things on my mind, no doubt, and also I can't really see the gauge all that well. They make them so small these days, don't they? In fact, I thought the temperature gauge was for fuel, and vice-versa, so had absolutely no concern about petrol until the car spluttered to a halt.

Such a silly thing to do, of course, but then, we all make mistakes. It makes life more interesting, they say. LOL.

Because of my earlier-than-expected departure from London, I have asked the BBC that there be no retirement 'do' at all. I know they always organise something for senior staff when they leave, but I think it's better this way – I've always hated parties anyway. Sorry to disappoint you all. I know how much some of you love to go dancing at the local discotheque!

May I take this opportunity to wish you all good luck for the future, which will no doubt be extremely challenging, but exciting, in our new digital, multi-platformed media age.

From my time at The Corporation, I know how special and talented you all are, and just how all the teams at the BBC can create really vibrant and diverse programmes which reflect a changing Britain, and bring to them the synergy and range not possible anywhere else.

I must confess that I have asked myself, on more than one occasion in recent years, when observing the changing media landscape and the profusion of digital and internet channels, 'Well, just what is the point?'

So I think that now is the time to retire. Because there has to be a point, doesn't there, if one wants to carry on in a career?

The BBC is special. Please take care of it.

And remember: problems are solutions waiting for happen, which is a big positive, going forward.

Good luck!

Regards,

Hugo

*

No Man is an Island

Rasmus remembers the rain.

He remembers when he stood alone on the roof of the building that night. The night he came alive with a bolt of lightning – the night he shed his skin – the night everything changed. It seems so long ago now, but it was just a few months in the past. His new future life has taken him so far already.

'So, it seems to be all change at the BBC,' says Rasmus, after announcing the departure of both Benjamin Cohen-Lewis and Hugo from The Corporation to the X-TV production team.

'About time too,' says Ravi, before correcting himself. 'I mean...the BBC should be, well, y'know, the BBC...'

'Yeah right,' says Gary Wu, sarky smug, 'that's genius, that is, innit?'

'What I mean is that when they started broadcasting – first radio in 1922, then TV in 1936 – they were the first in the world, and now...'

'Now,' yawns Sebastian, 'they're just sooooooo booooorrrrriiiiiinnnnngggg!'

'Oh totally,' says Anita.

'Oh yah,' says Mercy.

'Oh that is *so* true, yeah?' says Debs.

'But the competition should be strong,' says Ravi, thinking quickly, 'because that will make us stronger too.'

Thursday beams a knowing grin from where he stands by Rasmus – he likes quick-thinking in anyone, even his enemies. He remembers an Indian shopkeeper in his country who suggested that Thursday and his men could take the money and stock from his shop, if they agreed to drive the shopkeeper and his family to a secret hiding place, which happened to be in territory near the UN peacekeepers' base. Thursday grinned then as he is grinning now, when he agreed to the deal and promised to protect the shopkeeper and his family. Conveniently, the dead do not remember promises...

'People,' says Rasmus, 'Ravi is right.'

Ravi heaves an inner and silent sigh of relief.

Rasmus speaks:

'Strong competition is a good thing – it raises everyone's game. And when we take over the BBC, we don't want to take over a complete disaster zone, do we?'

So this was the plan – where X-TV was going all along. This was the destination of the journey they were on, to take over the BBC? None of the producers had known for sure about this before now.

'That is just well cool,' says Gary Wu, loving every moment of this audacity.

'Totally awesome!'

'Oh yah.'

'Absolutely,' says Ravi.

'But, before we can do that,' Rasmus says as the room descends into a hush, 'we have a great deal of work to do.'

He looks at Thursday and a slideshow presentation is projected onto a screen.

'The Island,' says Rasmus, 'will be our best show yet.'

Minty and the King

'Fantastic!' says Sir Peter Baztanza, pacing around the room like a hungry lion. 'I'm loving this!'

'Please, Peter, have a seat,' says Minty.

Peter Baztanza has never liked sitting down, and has certainly never liked to sit down when asked to by others. But, seeing as he is in the office of the new Director General of the BBC, and as he knows he has Minty Chisum where he wants her – namely in a corner and desperate – he does as he is asked.

Ten minutes later, the deal is done. Sir Peter Baztanza has agreed to work for the BBC on new and existing reality TV shows, as a 'consultant'. He has named his price. Two million pounds immediately, another two million after shows are broadcast, plus 50% of revenues and marketing for all new shows – plus international rights to them, with full rights reverting to his company, Grin TV, after five years, across all media platforms, including all internet and digital channels, and even those technologies and platforms not yet invented.

Minty knows it's a ball-breaking deal of a kind that no-one has ever managed to get from the BBC before. But she also knows that in this brave new age of broadcasting she really doesn't have a choice.

Sir Peter Baztanza is the king of reality TV – maybe the king of all TV – and has the track record and multi-million pound fortune to prove it. There is, quite simply, no-one as good as he is in consistently getting the ratings and producing hit shows, even if he does steal ideas from everywhere else. Of course, this is only what everyone in TV does anyway – but it's just that he does it so much better than everyone else. Always. And again and again.

Nothing he has ever been involved in has failed. Nothing.

There was never anyone in the TV world like the phenomenon that is Peter Baztanza – until now. Until Rasmus and X-TV emerged from nowhere to challenge his perfection.

But it wasn't as though Peter Baztanza was somehow seen as a rogue or lacking in respectability. He had, after all, got a knighthood from the Palace – apparently, his shows had some big royal fans. Indeed, he felt deep regret when Princess Diana died – she would have made a great host of reality TV, perhaps on a panel for 'PopStarzzz', or out doing some chat show in America. Diana's death had affected Peter Baztanza deeply. Just think of all the money they could have made together!

Minty, of course, now had to get the deal past the BBC Trust, where it would get final approval – and then Baztanza would be working for her, and X-TV would know what it was like to be under attack. Getting this approval would have been a problem in the past, but now she had the BBC Trust in her pocket – she'd taken steps to discover many a useful skeleton on BBC Trust members' cupboards. To all intents and purposes, Minty was now the new queen of all she surveyed, an all-powerful monarch, unlikely to be challenged. Invincible, even.

Any complaints were much more likely to come from the politicians – the usual meddlers and wafflers, who would compare Baztanza's deal to what a single mother in a council flat had to live on, or moan that to pay so much to an individual out of public funds was obscene. Well, maybe it was obscene, disgusting, vile, exploitative, unpleasant and outrageous - but that was TV, and it was clearly what the public wanted. The same public that watched Peter Baztanza's reality shows by the million; the same public who were abandoning the BBC to watch X-TV; the same public who would just go wild about whatever lowest-common-denominator reality TV trash was, at that very moment, fizzing into life in the brilliant brain of Peter Baztanza.

'Fantastic!' says Sir Peter Baztanza, his grin wide as a lie.

Minty, too, is optimistic. For the first time in goodness knows how long, Minty is working with someone who not only genuinely knows what they were doing in a nuts-and-bolts 'getting things done' way, but who also genuinely understands TV in all its terrible, wonderful, horrific beauty and who has a real feeling for what kind of programmes people want to watch.

She knows television is a wasteland - an invention that spews out crap twenty-four hours a day to the four corners of the world, that isolates people as well as making them feel they belong, that makes them happy and unhappy at the same time, that entertains by mocking and hurting and appealing to the worst side of human instinct.

No, Minty makes no excuses for television - she knows it's built on a steaming stinking cesspit of everything that is worst in human nature. She knows it's an apocalypse of inane stupidity and vanity, a regression to the primitive, basic caveman way of seeing - after centuries of civilisation that produced the greatness of the written word and the educated, cultured mind.

Yes, television is a wasteland alright, Minty has always thought – but at least it's *her* wasteland. It is what she knows. It is all she knows. And she knows she belongs in it, and always will.

'I'm loving this!' says Sir Peter, admiring his programme idea outlines for reality TV shows on the projector screen.

'Fuuuucking hell!' says Minty, wide-eyed at it all, her face smeared with a rare smile. 'This is fuuuucking faaaaantatic!'

It is traditional for the DG to meet the Prime Minister once appointed, and also to attend Downing Street functions. So it is agreed that to kill two birds with one stone. That evening, Minty will attend a Number Ten function welcoming a Chinese delegation and meet the PM for the first time too. Toby will accompany her.

The chauffeur – middle-aged, smart, silent – sits indifferent in the Bentley. When he sees Minty step out of her front door, he gets out, opens the car door, and closes it gently behind her. Then, almost silently, he glides the car away from the kerb and out into the thrust of London.

The journey from Hammersmith will not take long at this post-rush hour time of night – and Minty intends to enjoy every single tarmac-ed inch of it too, just as she intends to enjoy the spacious upholstered back seat of the chauffeur-driven car, and every other perk that comes her way. It has taken her thirty years to get where she is, and she intends to relish every single overprivileged minute of it.

Westminster is only around five miles away, and can take less than twenty minutes on a good day. But this is not a good day. The traffic is heavier than usual for that time in the evening, for whatever reason.

'Why the fuck can't I have a police escort then?' says Minty to Toby, who thumbs a memo into his smart phone to research later.

They cruise through Kensington. From there, it's just a short hop to Hyde Park Corner, Constitution Hill and the Westminster bubble.

The world shines wet and reflective on the pavements, the electric lights of shops windows mirrored thin in the puddles, and all looks peaceful, though there is definitely a certain something in the air. It is not anything definite, but just a feeling, an aura, a tense sense of something wrong – a twisted out-of-sorts nerve twitching within the city, a noticeable nervous tic under the skin.

As the car glides along, Minty and Toby look out at the people. There are the usual sad-faced commuters a-scuttling home, and the smiling tourists who always stand out because of their perpetual holiday cheerfulness. But there seem to be many loitering youths lining the route too. It has become a common sight in recent times - all these youngsters mooching about, unemployed and maybe unemployable, looking for any opportunity for trouble that reveals itself.

London never used to be like this - not *quite* like this anyway. These young kids look as though they've emerged from some strange underground land, crawled up and out into the daylight, a sinister and unwelcome underling infestation invading the streets of a divided, booming city. They know they are not welcome, and never will be, in the city they call home.

Minty looks out at their faces, and their faces stare back zombie-blank, then seem to contract - to clench like angry fists - at the sight of the brand-new Bentley. A resentful yearning look replaces the blank nihilism of before: a hunger, a want - maybe a need. These millennial kids have nothing, Minty knows, and they will never have anything either. There is just not enough 'something' to go round any more. Sorry, 'Generation Snowflake'! All gone! Just the way it is. Musn't grumble. Keep calm and carry on, and all that...The only time most of them will get to travel in a car anywhere near as expensive as the one Minty sits in now is when they're boxed stiff in the back of a hearse.

The traffic thickens and clogs as they approach Hyde Park Corner. Suddenly, a hand thuds on the window. Then another, on the other side. Faces pressed on glass. Eyes peering inside to where Minty and Toby sit.

'Don't worry, it's locked,' reassures the chauffeur, dialling the police on his mobile.

'Yobbos,' he sneers under his breath at the teenagers outside.

'Chavs,' spits Toby.

'Cunts,' snides Minty.

Dark young faces press against the car windows, squeezed into deformity by the glass.

There is nowhere for the car to go, sandwiched between other cars in this traffic jam. No escape. They just have to wait.

'You can draw the curtain, if you like,' suggests the chauffeur.

Curtains in cars is a novelty to both Minty and Toby. It had never occurred to either of them that they had curtains with which they could shut off the world, like on a plane. They simultaneously draw the drapes, though they can still see past the world through the windscreen in front.

The fist-banging has now become a rhythmical pounding drumming, as if calling a tribe to war against an old enemy. This, Minty thinks, must be how it feels to be a pervert in a prison van with the paedo-fixated public baying for blood outside.

'Oh no,' thinks Minty. 'Fuck NO!'

She can't be caught up in this - not in her first week as BBC Director General, not when she's on her way to Downing Street! She pulls back the curtain.

'What the fuck is wrong with this cunt of a country?' she shouts through the window glass at the faces outside - all ugly, alien, young.

Oddly, the cars in front do not seem to have youths gathering around them, and nor do the cars to either side. She looks through the back window. Again, the cars there seem to be unharassed.

'Old Bill'll be here soon,' says the chauffeur. 'Best stay in the car, miss.'

'Oh really?' sneers Minty. 'You don't fucking say.'

They watch as several hoodied youths - whose glazed eyes suggest they're skunked up or otherwise stoned - jump onto the bonnet of the Bentley and start jumping up and down. The thump of boots pummels the roof of the car.

Everywhere Minty looks, she sees a mob. It has come from nowhere and now seems, somehow, to be swallowing them like some strange alien blob of bacteria, a giant avalanche of anger enfolding them into its core.

'You're just jealous!' shouts Toby at a teenaged face which presses its first attempted moustache up against the car window – one which, rather disturbingly, looks a bit like a young Lenny Henry.

'The correct word to use is *envious,* Toby, not *jealous,* though no-one seems to know the difference any more. This is *not* fucking Othello!'

It is a reference that flies over Toby's head like a brick.

'They're just targeting the car, not us, Minty,' he says. '*Soooo* predictable.'

Sirens screech in the distance. Boots stop pounding the roof of the car. Then the stompers jump off the bonnet. Soon, and well before the police arrive – (as per usual!) – every single member of the mob who had attacked the car has vanished, melting away like mist into the anonymous evening streets.

'CCTV,' says the chauffeur. 'That'll nail 'em!'

'Let's just get the fuck to Downing Street, can we?' says Minty.

After giving brief details to a dim-looking police officer young enough to be his son, the Chauffeur obeys.

This time, they have a police escort.

*

Marked as CONFIDENTIAL
From: Minty Chisum, Director General
To: Oliver Allcock; Lucinda Lott-Owen; Sangeeta Sacranie-Patel
Cc: Penelope Plunch
Subject: Important Changes at the BBC

Thank you very much for all the good wishes received following my recent appointment to the post of Director General of the BBC.

As many of you know, I have been with The Corporation for my whole adult life, so I really do know what works and what doesn't. The fact of the matter is that over recent years the BBC has been focusing on the latter, rather than the former, and that is something that my leadership here will certainly aim to alter.

There are going to be some major changes soon, here at the BBC, and some of you will not be remaining with us. Dead wood will be cut out and mediocre managers will be expunged. In order to compete with the numerous television channels, especially those online, we need to constantly improve and evolve. If we do not, then the BBC has no future.

That is why I have secured the services of Sir Peter Baztanza, one of the leading producers in television today. I am pleased to say that we are already taking several of his ideas into production, and I am sure I can rely on the support and hard work of you all in making our vision of a successful and vibrant BBC a reality.

This is make or break time for us. We all know how X-TV has managed to become the most popular television channel in this country, seemingly overnight, and that we have to compete with them, despite our disadvantage in having to obey the laws of the land.

It may seem impossible for us to compete with such a company, whose wealth and power seem to grow exponentially. However, I have faith that we can not only compete with them, but also beat them in the ratings, so that the BBC becomes the creative and popular force it used to be, and which I am confident it can be again.

I look forward to our joint success in the near future.

Regards,

Minty Chisum

BBC Director General

*

It is said that to succeed in life, you need two things: ignorance and confidence.

The new Director General has both – but the BBC cannot and will not be successful. It is an impossible task, for several reasons.

The BBC has its hand tied behind its back by regulation, and is stifled by lack of ambition and an unwillingness to take risks. True, Minty has hired Peter Baztanza – but then again, does she really think that this will save her? It is wishful thinking forged into an artform.

Minty Chisum has absolutely no idea how the world is changing – just as she has no idea that all information at the BBC is available to those who know how to access it, that any email she sends or phone call she makes is shared with her competition. And who's to say conversations are private anyway? Certainly, the technology exists to listen to everything. Privacy, as we used to understand it, no longer exists. There are no secrets any more.

Minty Chisum also does not appreciate one little inconvenient truth: we have the best government money can buy, and more than enough finance and freedom to make whatever programmes we want, when we want, how we want.

Every man needs an enemy; every woman, maybe more so.

Minty knows that better than anyone – but if I am to be her enemy, it is nothing less than the role I created for myself by creating X-TV and changing the game.

I am Rasmus, and this is my reality.

*

Calvin is chewing a large mouthful of one of the sausage rolls his mum has made for him. It's his first day back after a weekend at home – a weekend when he chose to 'come out' to his mum and his nan about his new career presenting shows on X-TV, most of which seem to them like the dirtiest filthiest television programmes they have ever seen.

'Iss just a temporary thing like, mum, int'it? A first *rug* of the ladder, like...'

His mum nods sweetly and goes out to the kitchen to put the kettle on.

'Even Madonna started in porn films! And I'm not actually in them anyway like...so I'm not doing nowt really... an't money's reet good...an s'not not like I'm actually *doing it* on't telly...'

Calvin suddenly realises he's alone and talking to himself. And he knows right there and then at that moment that his mum will never accept – can never accept – his new career, no matter how many shoes and handbags he buys her.

He even got her a new widescreen TV, but it remains there in its box, like some shackled mechanical monster. His mum says that she prefers her old telly anyway, that there's "nowt wrong wi'it", and anyway, she doesn't watch much TV these days.

'Radio's much better, love,' she always says. 'Got much *nicer* pictures and that, when you think about it, int'it?'

When Calvin thinks of the things he has seen while presenting X-TV shows like 'Granny Gangbang' and 'Dwarf Orgy', and all the 'Celebrity Suck-off Specials', he can't help feeling that she may have a point.

He was, he knew, ashamed of what he was doing. But what was the alternative for someone like him?

It's only a temporary thing, he kept telling himself, until something better turns up, and he can get back to being a pop star, and singing, and making records, and being famous –

for something good, for his talent, and not just for presenting pervy porno on't telly.

Now, back in London, Calvin remembers his mum as he sits scoffing a sausage roll whilst watching, on the monitor, an eighty year old woman, (a former bit-part actress who has become an unexpected celebrity in her twilight years thanks to her gutsy performances on 'Granny Gangbang'), chewing the bulbous cock of a nineteen-year old former youth football star who had to retire through injury - possibly a groin strain, thinks Calvin. The footballer found fame, or at least D-List celebrity, for being sucked off under the duvet by two former models on the goldfish-bowl reality TV show 'Watching Us Watching You', where people - (mostly young, hormonal, and staggeringly thick people) - are locked in a house together for weeks, supposedly as a social experiment, but really actually just to try and get them to have sex together.

At least, Calvin thinks, we don't actually have to be on set with them - all these 'former' people who now do this on TV for a living. I mean, what kind of a job do you think this is?

The porno shoots are always done at studios to the west of London - or sometimes overseas, in Eastern Europe, if the legality is dubious. Since the birth of the Internet, however, the UK authorities have not really even bothered to enforce any available anti-pornography laws that exist - because the instant online availability of vast quantities of such material would make any such attempt as absurd as it was pointless.

'Fame is the thirst of youth,' sighs Sebastian, watching a piston of a penis mechanically pounding the face of an old lady who bears an uncanny resemblance to June Whitfield, 'except, perhaps, for grandma there...'

Calvin thinks of his own nan - so sweet, kind, gentle. He couldn't imagine her even having sex at all - though, he supposes, she must have done, at least twice, to have two kids. But the thought of her doing what that old lady was doing on the monitor in front of him was more than disgusting and perverse and sick – it was wicked. Cruel and wicked. Evil, even. It was just plain out of order. It was, in a word, *wrong*.

By coincidence, Calvin swallows a lump of his sausage roll just as the old granny on the screen is swallowing the first of the copious emissions from the footballer's balls, which sort of puts him off having another bite.

'Sure you don't want one?' says Calvin to Sebastian, his producer for all the sex shows, which he now presents alone, Alicia having been assigned to one of the soon-to-be-premiered programmes. 'Me mum makes 'em for me, like...'

Sebastian smiles at the sweet innocence of this cute boy-man presenter.

'Ah Mrs Snow, how ye have been blessed...'

'You what?' says Calvin, flakes of pastry scattering with his words, as he takes another bite of his sausage roll.

'Naughty naughty,' says Sebastian. 'Talking with your mouth full!'

'Sorry,' says Calvin, as automatically as he would to his mother.

They both turn to the monitor where the granny's mouth is still uncomfortably full of male member. She is definitely trying to say something, though all anyone can hear is a sort of groaning grunt.

'Want some?' says Sebastian, holding out a wrap of paper with some white powder nestling in its centre.

Calvin shakes his head, almost blushing. He can put up with watching all the dirty disgusting pervy sex – just – but he certainly doesn't want to get involved in drugs of any kind. It was bad enough being a pervert – even at one stage removed – but if he became a druggie it'd surely break his mum's heart.

He'll stick with the sausage rolls for now – for his mum's sake, at least.

And X-TV's, of course.

Rasmus reads the email - the highly confidential one Minty Chisum sent to her BBC colleagues. He is sitting in the soft glow of perfect lighting and Tchaikovsky at the house in Holland Park. Thursday enters carrying a tray, on which tinkles exquisite antique porcelain. It is time for tea.

The small certain smile on Rasmus's face says more than words ever could. He is winning, and he knows it. What amazes him is that the BBC, with all its money and power, can't work out how easy it is for outsiders to crack its security systems. Probably many a teenager with a laptop could hack into their emails, and the only reason they haven't, no doubt, is because of the relentless bureaucratic tedium of the information the messages are bound to contain.

One reason X-TV is becoming so rich and powerful is because of the information it gathers on all those watching X-TV online - names, ages, email addresses. That information is highly valuable to advertisers, who can mine that data and then target their personalised adverts far more effectively at internet-savvy young consumers, thus both saving and making more money. That is why they will pay huge sums for it, in addition to giving X-TV a cut of their profits.

In contrast to the complex information technology needed to track users - a task carried out by the specialists and IT engineers on the lower floors of the X-TV building - getting access to emails is easy. Rasmus has read every single email Minty has sent since the founding of X-TV, as well as all the BBC emails his staff feel might be of interest.

Rasmus is surprised - and amused - that Minty seems to be under the impression that X-TV is somehow worried about the new BBC season of shows created by Sir Peter Baztanza. The fact that X-TV can bypass all UK and European laws by filming overseas and then broadcast only via the internet directly into the UK - and indeed all over the world - seems to have largely escaped the BBC's notice. What this essentially means is that anything the BBC does, X-TV can do better, and make more extreme, with more sex and violence – and, yes, death. The BBC's hands are tied – and not even the brilliance of Peter Baztanza can work miracles.

'Laws are like cobwebs, which catch small flies but which let wasps and hornets break through,' Rasmus says to Thursday. It is so easy to bypass the law – break through it, ignore it completely – if you know how, if you have enough money, and if you have no fear.

Lack of money is surely the root of all evil, as Thursday well knows.

As it is, X-TV's regular shows are still dominating the ratings – and that's even before the planned future programmes will increase them, which Rasmus knows they surely will.

For now, the usual 'Granny Gangbang' and 'Dwarf Orgy', and all the other shows which make sexual perversion essential viewing for the curious and/or horny, are keeping the public – of all ages – focused on X-TV.

People are interested in sex – it is basic, it is human, it is profitable. That's why it's on TV, on all channels, and everywhere else too – in magazines and adverts, and oozing and dripping off millions of computer screens every hour of every single day and night all over the world.

The competition – both the BBC or independent channels – is hardly worrying or in any way inspiring. The usual soaps and magazine shows, cooking and property porn, sad little quiz shows, tedious tired talent contests and endless crawling news – cheap, salacious, melodramatic, exploitative, dumbed-down bite-sized news, with presenters waving their arms around like windmills as the headlines ask the usual meaningless scaremongering questions, chief amongst which seems always to be 'Are Your Children Safe?'

'Of course they are,' Rasmus says as Thursday hands him his cup of tea, a little milk and no sugar. 'They're all at home watching X-TV.'

'Cool,' says Calvin to camera, his boyish smile innocent and open but absolutely fake too. 'That's reet good – a right laff, int'it – and we hope you can all join us tomorrow night, here on X-TV, like, for more 'Anal Adventures' and 'Celebrity Suck-off Specials'. So see yers all then, like.'

The camera light goes off – Calvin is off the air at last.

'Good boy,' says Sebastian. 'I bet that right now, even as we speak there are simply millions of sad sods wanking themselves raw over your pretty little angel eyes.'

'Oh shurrup, can yer?' Calvin says.

Sebastian notices the change. Calvin usually just blushes when Sebastian starts a bit of dirty filthy flirting. But now he seems genuinely upset.

'Why, what's the matter, dear boy?' says Sebastian.

'It's nowt,' says Calvin, 'it's nothing, really.' Sebastian looks at the hurt in the boy's eyes and sighs. He has seen it all before.

'Look, Calvin, dearest, it's only TV,' says Sebastian, 'and that means at the end of the day, it's all just a perverted pile of shit – that's just the way it is these days.'

'But why?' asks Calvin.

My god, thinks Sebastian, this is serious: the boy's actually using his brain – he's actually thinking! Dangerous – very dangerous indeed, especially in TV-land.

'Because...that is what people want to watch on TV. You want to give the people what they want, don't you? Be popular? And famous? And rich?'

'Yeah, I mean, no...' Calvin pauses, then starts again: 'I mean, yeah, course I want all that stuff, but for being a singer, not for presenting porno, like...'

'Ah but without pornography the world would be a much worse place – we wouldn't have all our wonderful new technology for a start. Just think of all the things men invented just so they could watch disgusting, perverted filth: the camera, the video, the DVD...'

'But, y'know, I sometimes think – sort of, whass the point?'

Sebastian jumps to his feet as though Calvin had lobbed a grenade into the control room. He grabs Calvin's tightly-muscled, almost-hairless forearms.

'Never, ever, ask that question,' he says, staring into the boy's limpid blue eyes, wishing he could dive into them forever, just like they do in the gay porn novels he reads to put himself to sleep.

'But why?'

'And never ask that question either. Trust me – it'll drive you mad if you let it. Just remember that you're providing a service to the millions out there who love what you do – and you're good at it too, boy!'

'But I don't want to be good at presenting porno, I want to be good at singin' and dancin' an'...'

'Yes, and so this is a first step...Scuse the pun.'

Calvin isn't sure what a pun is, so ignores the last bit.

'Popstarzzz was the first step,' he says.

Sebastian remembers Calvin from the show. He knows that being famous – even a little bit famous – and then having to return to a life of ordinary and boring anonymity, especially with no money, but with the added burden of being famous for having been ridiculed on national TV, would be enough to break most people, but not Calvin. That's how Sebastian knows this kid is tough – just like him. Just like anyone who works in TV – and stays.

How all the people who know them must howl in laughter at their public humiliation, even for the ones who can actually sing, like Calvin. How hard must that be to cope with, especially for one so young. Reality TV could be a brutal and unforgiving teacher indeed – it could, and did, indeed destroy all it touched.

'So this – X-TV – Celebrity Suck-off Special and Granny Gang Bang – this is the second step then, another step on the yellow brick road to stardom, Dorothy. Even if it does involve some very disgusting things being done to Toto now and again, and maybe to a Munchkin or three as well.'

Calvin turns away, not bothering to hide his sadness now the cameras are off. He only looks happy these days when he's in front of a TV camera.

'It could be worse, dear boy,' says Sebastian. 'You could be selling cheap crap on the shopping channels, or doing ads for acne creams, or...'

'Goodnight Sebastian.'

'Goodnight, sweet prince,' Sebastian says, but when he turns around he sees that Calvin has already left, fleeing outside to the car that will drive him home - or at least to the impersonal luxury apartment that X-TV provides for him in a West London mansion block.

Later that evening, Calvin calls his mum and thanks her for the sausage rolls, made like only she can make them, and which, he says, he and his producer have shared and enjoyed with some cans of pop. He makes it sound as though they were enjoying a cheerfully boyish midnight feast at cub camp. He omits to mention that they scoffed them while watching a toothless pensioner fellate a teenager on a large HD television monitor.

Calvin tells his mum he loves her, puts down the phone and bursts into tears. Then he cries himself to sleep, alone. Again.

Rasmus speaks

Rasmus is in the office on the fourteenth floor of the X-TV building in Hammersmith. Thursday stands by his side.

'Thank you for your continued support,' says Rasmus, to the face on the screen - the Chinese face of one of X-TV's valued international partners.

Thursday thinks of how far they have come - of how far *he* has come. He knows how fast things can happen. He knows that bigger things will happen very soon.

Rasmus opens a file and reads. A photograph of Minty

Chisum is there, together with the BBC team of Toby, Oliver, Lucinda, Sangeeta – and, finally, a photo of Hugo which has a big black cross marker-penned across it. Rasmus pulls Hugo's photo out of the files and then into a second file. He is no longer relevant to X-TV. The others, however, are.

'And they still don't know...' says Rasmus, partly to himself. Thursday smirks.

Outside, the rain whips against the windows, driven by a wild westerly wind. Thursday looks out at it, blankly. August weather in England: high summer.

Rasmus nods at Thursday who leaves to prepare some tea. The Gillray prints stare down sneering at Rasmus in the room, the only observers of everything, an audience of mocking, meddlesome eyes.

The screen on the desk then switches to video-conferencing mode once again, and a face appears on the screen.

Rasmus smiles back at the face, and they talk.

Hugo's Welcome to Wales

The last day of August. Hugo sits in his garden in Llandoss – in a small roofed area at the back of his cottage, on account of the relentless Welsh rain – a glass of wine in his hand, remembering all the happy times he spent working in television – such as that day, all those years ago, when Princess Diana died in Paris. It's the anniversary today.

'Happy days,' thinks Hugo, his mind meandering back to days that were not, he knows, happy at all – not until they'd been through the magical memory machine in his skull and been reprocessed and reshaped, like so many chunder-churned chicken nuggets, into pleasant and palatable memories that are not quite so dreadfully desperate and depressing as the stark reality.

In his heart, Hugo knows the memories are not happy ones – but where's the harm in pretending? He knows that faking it and pretending to be happy, especially when completely and utterly sloshed, is the nearest he's ever going to get – so why not just enjoy the fantasy to avoid the reality? Isn't that just really what most people do all the time anyway?

But even the real memories – the bad memories – were certainly not as unremittingly awful as his present reality was turning out to be. He thinks that perhaps it might have been a mistake to leave his home and his job in London, after all.

The village nestling in the valleys, with the view of a grey-green hill – which, Hugo had recently learnt, had been dubbed 'suicide mountain' on account of it being a favourite spot for 'jumpers' – had looked so lovely when he'd seen the photos. It looked pretty too on the one hour trip he'd made on a balmy sunny day in early summer – one of as many as twenty rain-free days every year in Llandoss, the locals later inform him.

There was nothing really wrong with the village – it was just a few rows of terraced houses, with a few cottages, a pub, a corner shop – but appearances were deceptive. Like many mining villages that had sprouted from the soil to house the workers of an industrial past, this village no longer had a mine. Its purpose had gone, but it was still there, a vestigial organ rotting away, useless and without purpose. It still had the people, however, around three quarters of whom were unemployed, and a significant proportion of whom seemed to be wasting away through drink or drugs.

To a cultured, educated man like Hugo, it sometimes seemed as if he'd parachuted down onto some weird incomprehensible planet inhabited by strange alien creatures – who did, it's true, speak English, though with such a broad Welsh Valleys accent that Hugo struggled to understand what most of them were saying. Hugo holds up his drink and stares deeply into its rich ruby-dark oblivion, a scarlet stained glass window into his bruised and bloody soul:

'Something, um, tells me, Toto, that we're not in, um, Kansas anymore,' Hugo slurs. 'Or Twickenham, for that matter...'

The briefest of chortled laughs and he empties his wine glass.

Hugo really did sometimes feel as if he'd been washed up on some strange island – though he was only just over 200 miles from London. It was just all so *different*. Hugo understood that. He understood that he didn't understand the culture or the people here – he could see that there was such a divide between them and him that he doubted if he'd ever make more than small-talk about the weather with any local. His RP English accent certainly didn't help either. He expected they'd have got on well if they'd met elsewhere, in other circumstances, in another time and place – but they hadn't, so they didn't. It wasn't their fault. It wasn't his fault. It just *was*.

After just a few weeks living in the cottage, the isolation of his new life became clear. And he used to think that life with Felicity in Twickenham was lonely...

Hugo fills his glass again, ignores the spills, and drinks a toast:

'To, um, Diana,' he says, 'may God, um, bless her, um, and all who sail in her', says Hugo, his mind muddled, befuddled, and most probably pissy-puddled, ever since he had started the day's second bottle of red. Or was it the third?

'Probably a ship somewhere called that anyway', continues Hugo's addled thought processes. 'Princess Di...who died... there's prob'ly something somewhere called everything, if you look hard enough, probably something or someone else called Hugo Seymour-Smiles too, maybe now or in the past or the future, or even in an alternative universe which is exactly like ours, with the same people and everything, but different, or maybe this is all a dream anyway, this cottage, this rain, this glass of grape, this England, though not England, but Wales, where they speak English, mostly, sort of, and the Tudors were Welsh too, and you can't get more English than them, but anyway, off the point, back to alternative reality, like in one of those tricksy postmodern Hollywood films that think they're being clever by recycling crass clichés from science fiction, never was one for science fiction really though, or maybe this is just one of many realities, reality being a subjective perception, or maybe we're all in some giant game, with the gods moving us like chess pieces, just like in Jason and the Argonauts, that's my favourite film, y'know, what with the Ray Harryhausen special effects, especially the skeletons, who keep on coming, unable to be killed because they're dead already, and are all just skeletons, so Sinbad's sword goes right through their ribcages, but not Sinbad's, it's Jason's, just getting a *mit bixed* up, I mean a bit mixed up, David Jason got his name from that film, not David, but Jason, you plonker, nice plonk this, Australian, better than the French, more consistent, stronger, gets one pipsy and tissed sooner, tipsy and pissed, when will this rain stop, will it ever stop, will it never stop, ever, never, ever, never....'

Hugo empties his glass, yawns at the rain and makes a decision. He realises that alcohol is not the answer. No, he also needs TV. Not *wants* – *needs*.

He really, *really* needs it. Needs it as much as food or water, or booze. Not having it, not bathing in regular televisual electric warmth, has left Hugo feeling as bad as any addict deprived of their fix. He hates himself for it, but he has to admit that TV has defeated him, got its lurid luminous hooks into him forever. There is now just no way he can ever live without it.

He realises, at that moment, right there and then, while watching the wet Welsh rain running rivulets down his garden path, that television was, and is, his best – his only – real true friend.

He remembers when he was about to leave London, how he decided to take a walk one evening around the area where he and his family had lived for so long. Dusk was falling as he wandered through the suburban streets, hoping to see the sights of his youth through a thousand suburban front windows: children playing board games or doing their homework, fathers smoking pipes and reading the paper, mothers bringing in dinner to the family at the table. But what he saw had a dreadful uniformity to it: through the window of every single house shone the electric blue glow of the television set. And, in the houses with no net curtains or closed blinds, where he could actually see in, all he could see was the occupants sitting on sofas, gazing lazily at the hazy blue hearth, no doubt all watching the same celebrities on the same shows, all basking in the light of a gigantic omniscient, omnipresent, omnipotent TV sun-god. And it was even worse with computers and the internet...

Hugo rubs his head. It was bandaged at the hospital following a fall – at least, that's what he told the nurses. What actually happened is that Hugo had gone, not for the first time, to the local pub, where he had drunk several pints alone.

It was when Hugo was halfway home that it happened. Heavy footsteps suddenly rushing towards him, a large force thumped on his head and then he felt the pavement flat on his face. They didn't even bother to steal his wallet.

The fact that Hugo had not been having a very successful time in the Valleys was especially brought home to him by the failure of his organic crops. The rats had eaten all his seed potatoes and the constant rain had seemingly drowned a lot of his other plants – except, he noticed, the Japanese knotweed which was going strong, so much so that it had actually caused a large crack to appear in his garden wall. It would be fair to say that he was now at a low point.

This was especially so as Felicity's enormous and punitive divorce settlement meant that he would really have to rely on the crops he grew for food, and not just be a hobby gardener.

He only had about six months' worth of money left. Then he'd just have to rely on his income from teaching at the university. Except that he had no income from teaching at the university any more because he had resigned – and would have been sacked if he hadn't.

The final straw was when his departmental manager ordered him to pass students on courses that they had clearly failed, in order to meet targets, and thus not put off international cash cow students who wanted to get a good degree when they paid many thousands to come to a UK university. This was called 'good practice', apparently, and everyone was at it in every university in the land. Hugo walked out of that dodgy degree factory the same day. Well, you have to hang on to at least some self-respect, don't you?

He could have read a self-help book – (after all, he was a direct descendant on his father's side of the pioneer of self-help books, Samuel Smiles) – but they were not really his thing. No – he knew exactly what he needed: he needed television.

Hugo's self-help, however, was not much in evidence when he badly misjudged the next step on the stairs when making his way down them with the TV set, fresh from the loft, in his hands. Neither was it apparent when he came to at the bottom of the stairs with a splinter of plastic sticking out from his groin, where it had pierced his scrotum in a manner not dissimilar to that of a shish kebab skewer piercing a chunk of gristly meat.

As Hugo explained to the ambulance-men when they arrived, he had tripped and fallen, and as he only had the sight in one eye, it was so hard to judge distances.

They said nothing. They could smell the booze on his breath – as they could with most of their 'passengers' – and had heard so many excuses for embarrassing injuries that they barely bothered listening to all the lies any more. People were perverts. End of.

'Wonder how much a new TV costs these days,' thought Hugo in the ambulance, thinking it lucky that he should suffer such an injury *after* he'd fathered children, but annoyed that he had now broken the only TV set he had, which would mean yet another significant expense to come out of his ever-dwindling kitty.

*

All men would be tyrants if they could – and some women too.

To have power and to use it to achieve dreams: is this not the ambition of every human individual? The dreams may vary, as do many of the definitions of power, but the truism remains.

Every great dream begins with a dreamer – a creator – a beginning. I, Rasmus, am that dreamer, and I now make that dream reality.

The dreamers of the day are dangerous men, for they may act their dream with open eyes to make it possible.

The BBC and Minty want to achieve their dreams too, and their dream is for things to continue the way they have done for decades. But they too know that the past is over and the future belongs to others. In their hearts, they know they are fighting a losing battle.

He who fears being conquered is sure of defeat. That is why the BBC will fail and X-TV will succeed: we have no fear. They do.

I am Rasmus, and this is my reality.

*

Autumn Season

August became September, and in the world of TV, that means only one thing: the new autumn season and a time to launch new shows.

Minty marches down a corridor, fresh from yelling at some terrified BBC operatives. The BBC building is full of corridors which act as the veins of a monstrous machine-creature, carrying creative corpuscles to its heart – feeding it, enriching it, making it *live*. In theory, at least. But, as Minty marches along them, Toby in tow, she can't help thinking that they resemble the corridors of a rather massive and peculiar mental home.

Minty feels her heart beating fast and tight; her nerves are alive, fizzing like fireworks. This, she knows, is something called 'stress', which costs the BBC millions in sickness payments and early retirement every year, though it's never something

Minty has ever moaned about – unlike the contestants she sees on reality TV who always seem to be whingeing about it.

'Stress?' thinks Minty. 'None of this is stress – not real stress. Being in a war is stress; being raped and torn as a child is stress; living in a place that feels to you like hell on earth and thinking you'll never escape, that is stress!'

This last scenario is one that Minty has known all too well.

'But this? This is TV. It is not stress at all. It is fun. Apparently.'

But Minty is not having fun. No, because Minty is, in fact, very stressed indeed, and she tries to lessen her feelings of stress by doing what every other stressed person in charge of TV always does – shouting and swearing at others, and blaming them when anything goes wrong. And she always feels one heck of a lot better for it too.

Today is a big day for the BBC – perhaps the biggest ever. It really is make or break time for Minty Chisum too. The first of the new season of shows – 'Baby Love' – airs tonight. If it succeeds, and that success is built on, then Minty will go down in history as the Director General who saved the BBC. If it fails, then she'll just be remembered for taking it further down into the pit of failure and despair where everyone knows it has been languishing painfully for years, like a lame animal waiting for the slaughterman's stun-gun.

In her office, Minty has meetings with various BBC staff, co-ordinating everything, a spider at the centre of a web, twitching the threads – and all to catch those 'customers', the viewers, whose ratings they so desperately need. How can they fail, with Peter Baztanza on side? He has never had a flop. Never. And now he is on Minty's side and works for the BBC – and his shows will lift the Corporation to the status it used to enjoy, before all the digital channels and X-TV came along and spoilt the party. The BBC deserves to be at the top: it is its duty to be there – its right, even.

'Fantastic!' says Sir Peter Baztanza, who is pacing up and down like a panther.

'Please, Peter, sit,' says Minty, pointing to the seat opposite her. It is an order, not a request.

Momentarily, he stops grinning - an unusual thing for him - but then sees the look of desperate brittle tension in Minty's eyes, and complies. He does not want to provoke her notorious temper, doesn't want the distraction of a petty argument.

Peter sits. Silence. Stares. Waiting. Breathing.

'Ooo isn't it exciting,' says Toby, trying to break the ice.

Minty has decided to watch the first of the new season of shows here, in her office, with Peter Baztanza and Toby, on the flat-screen TV that hangs on the wall, like art.

She could have hired out a viewing room to watch the show on a huge screen with supportive, lickspittle BBC staff smiling from floor to ceiling and gushing praise about how brilliant and incredible and exciting and vibrant and cool they find the show.

But no - she is Director General of The British Broadcasting Corporation, and this is her office, so this is where she will watch the show - where she, the captain of the ship, will sail it, and all upon it, onwards into the unknown, hoping to reach a happy land rather than an iceberg, but ready to accept whatever fate lies ahead.

It is almost time. Minty allows herself a drink. The gin stills and calms the jangling nerves within her, and to some degree clears her brain of the tiredness, the confusion, the stress, the past, everything. 'What will be, will be,' she thinks, somewhat counter to her usual philosophy of personal responsibility. 'It's in the lap of the gods now,' even though she does not believe in any. She has done all she can.

And so, as the seconds count down to 7pm, Minty Chisum, Director General of the BBC, sucks in her nerves and watches a television screen, just like millions of others all over the country, most of whom she has nothing in common with whatsoever, and the majority of whom she utterly despises.

'Isn't he just gorgeous?' trills Tanya Tring.

Minty cringes – if there's one thing she hates more than gormless TV presenters, it's babies.

The studio audience coos, cameras panning their adoration of the blue-eyed baby boy gurgling before them.

'Sadly,' says Dr Pixie Fister, all-round baby expert and daytime TV doctor, 'little Ben is not well at all and is not expected to survive into adulthood.'

Tanya bites her lip and blinks her moistening eyes. Looks of concern on worried faces in the audience. More bitten lips. Hands to mouths. Hugs. Emoting.

'Let's...look at the film,' she says, fighting back tear-stick television tears.

The huge screen in the studio now shows the film about little Ben, to the inevitable minor key strains of that Michael Jackson song about a rat.

Tanya Tring is not sure what disease little Ben has except that it's not infectious and it will kill him before adulthood. This is great. A healthy baby is good, yes – but a baby with some awful condition or syndrome, especially one that will prove fatal, is television gold.

The film about little Ben gushes on. Tanya Tring and Dr Pixie Fister pretend to be watching it for the first time like the audience, though they have had to sit through it at least ten times already. 'Emotional make-up' – that's what the producer called it. Or was it 'fake-up'?

The film ends with a heart-rending hush. The cameras pan across the audience, with close-ups on the ones who are gushing their sobs most theatrically. All of the audience members look upset and moved, many in tears. Some cover their faces with their hands at the tragedy of a little life so soon to end. Perfect.

Everyone knows that pain makes better telly than joy.

'So, now,' Tanya says to the audience, but speaking soft and low, as to a lover, 'at last...the moment has come...'

The audience holds its breath. Tanya waits until she has them all, every one. And then:

'...to meet the best and most beautiful baby in the whole wide world – come on out Ben!'

Happy music. Cheers. Clapping. Lights. The sliding doors of the studio-set glide open and a woman walks out carrying a bundle.

The woman – Ben's mother, who looks suitably meek and vulnerable, and not too chavvy and fat (thank goodness, thinks Tanya) – sits with Ben, tiny and tragic, on the gay-rainbow of a studio sofa.

'Actually, the BBC is doing more than that,' says Tanya, as she turns to camera. 'Baby Love is giving you, the Great British Public, the opportunity to sponsor Ben, the first of our featured babies tonight.'

'Yes, and that means,' says Dr Pixie, 'that the lucky winners of our sponsorship auction can be assured that all the money paid will go direct to little Ben, for his treatment and care.'

'They'll get a fully-authenticated certificate of sponsorship too, with regular photos and updates on Ben's progress.'

'And, subject to the usual checks, the sponsor will have the chance to actually visit Ben, and his mum, at home.'

'The phone lines are open,' says Tanya, 'and the internet too, so be ready to place your bids. Your time...starts...now!'

Suddenly, at the foot of the huge screen and every TV at home, a counter shows the bids coming in. Soon, it reaches £50,000 and is still rising.

Ben is the first of many babies to get through that evening, which is why the show runs for a full two and a half hours of the most shamelessly mawkish emotional manipulation that television has ever witnessed.

The show ends as it began, with little Ben's sweet little terminally ill face gurgling at the audience in the studio as he leads all of the other available babies out for a final presentation for the cameras.

There is an almost religious cathartic glow on the faces of

those who have been present that evening – a feeling of sorrow and gladness and love which has truly penetrated their minds and squeezed every last drop of human emotion out of them.

But there is really only one thing on the mind of Tanya and Dr Pixie – and Minty Chisum. The ratings.

It is always all about the ratings.

Minty is in her office with Toby and Peter Baztanza, and now also with Oliver Allcock, BBC1 Controller.

'Incredible show, Minty,' he says. 'Just incredibly good TV.'

'Fantastic,' says Peter Baztanza. 'I'm loving it.'

'Of course you are, Peter – you created it,' says Minty.

Minty leans towards the laptop screen. Peter and Oliver stand behind her.

Tense silence. No-one breathes. Freeze frame. Pause button made real.

The instant ratings will not be entirely accurate, but they will give some idea of whether the show has succeeded, and usually a pretty good one, based on just a few sample households.

As she takes in the ratings, Minty's face loses its tense worried expression in an instant, the sunshine of cheerfulness rising in the smile that replaces it. For once – and for the first time in a very long time – the BBC ratings are good. Not outstanding, but good enough – which is perhaps not surprising given the vast sum the BBC has spent on advertising and promoting the show.

'Well done, Minty,' says Toby.

'Incredibly well done,' says Oliver.

'Fantastic!' says Peter Baztanza.

'I'm fuuuuuucking loving this!' Minty yells at the computer screen.

'X-TV will be incredibly disappointed,' say Oliver.

Minty says nothing. Nor does Toby. They silently wonder why X-TV did not try and take on BBC that evening and instead just show its usual sex shows. They must be losing their touch...

'The BBC is back! The BBC is back!' parrots Toby as he cracks open the bottle of champagne he has squirrelled away for just such an occasion.

I am Detective

The next day, Minty meets a man. He has been recommended to her by a contact at the BBC Trust, a former MI5 official who knows how to *get things done.*

Toby looks at the man. He is forty-ish, bearded, average-looking. He could easily have been just another member of the public in a boring job, a geography teacher at a failing comprehensive, or some jobsworth at the local council.

Minty too sees a face that is easy to ignore, that doesn't stand out, that won't get noticed. Perfect for a private detective.

'So, can you do it?' she says.

'Yes, I can,' says the man. 'In time.'

'How much time?'

'As long as it takes.'

The man stares blankly back at Minty's glare. She knows he is good, that he can unearth everything she needs to know about Rasmus, maybe find out who he is. Minty nods acceptance.

'I'll start straight away,' the man says. He leaves the office without another word.

'Not sure about him,' says Toby, 'seems a bit shifty.'

Minty says nothing. She stares out of the window, across West London, to where Rasmus sits in his X-TV office, alone.

Gladiator!

Later, Minty stares at the TV screen. Over in Holland Park, Rasmus watches too, Thursday by his side.

It is Saturday night and the second of Peter Baztanza's BBC shows is on air.

'Kill! Kill! Kill!'

The voice of the people has spoken.

On the screen, a masked gladiator, dressed in a bright blue plastic costume, and holding what looks like a trident, stands over the prostrate and disarmed red gladiator on the padded floor of the 'combat zone'. Lights flash and twirl, music thumps and crescendoes, before dying to a low expectant background hum.

A tension-tickling pause, then an almost unfeasibly deep movie-trailer American voice booms over the sound system:

'The winner...of this bout...and new Gladiator finalist...is... Vortex!'

The word 'VORTEX' punches itself onto the screen like a fist, then 'WINNER', slams like a bone-crunching kick over it. The screen graphic representing the opponent's name – 'RED DEVIL' – collapses like a building being demolished into a pile of cartoon rubble: a CGI broom sweeps it away. Vortex raises her arm in salute and acknowledgement.

Screams, cheers, yells from the crowd – the usual whooping and hollering deranged mental patient histrionics expected from a TV audience.

'Live...or die?' booms the American movie-trailer voice, (which actually belongs to a one-time minicab driver from Walthamstow called Wayne).

The presenters walk forward: a huge black man and former gladiator called Troy, and a tiny bird-like blonde girl called Jade who looks like she'd be crushed to dust if Troy so much as brushed against her.

Jade Joystick once represented Scotland in gymnastics at the Paralympics – and, although she came last, that led to a presenting job on CBBC, her lack of an arm being a definite plus due to the BBC's positive action policies, especially as she promised never to wear her prosthetic limb. The BBC insisted that the viewers would have to see her 'uniqueness', or as

Jade called it, her 'stump'. She always knew she would one day be grateful that her arm had been shredded by the family Staffordshire pit-bull cross, Buffy, when she was a toddler.

'Time...' the presenters mouth to each other, '...to vote!'

The public vote: that lucrative mock-democracy beloved of TV shows the world over.

The voting lasts one minute exactly, with deliberately tense and irritating computer-generated heartbeat music pumped into the studio gradually getting ever louder as the seconds tick-tock by. The studio cameras scan members of the audience um-ing and ah-ing about who to vote for on their keypads. There are only two options: 'Live' or 'Die'. We see one person vote for 'Live' and another vote 'Die' – it must be close! Then it ends.

'Voting is over,' booms the unseen Wayne, in his best Don LaFontaine accent.

Yells, screams, noise from the audience; music, lights, above and all around.

'So tell us, computer, will Red Devil live or die?'

Silence. The delicious lingering lust of a live television pause.

And then:

'DIE!' slams onto the screen, flashing manically as thumping booming music fills every space in the arena and the lights flash crazily all around. The percentage who voted 'Die' is 97.4%.

All members of the studio audience are now on their feet – and stamping them – and they chant 'DIE! DIE! DIE!' so loudly that it threatens to drown out the manic music.

Troy and Jade watch from the stage, each feeling like a Nuremburg rally Nazi Führer, but different, what with one of them being a buffed up black man and the other female, Scottish and armless, though she did, when a keen and sporty young teenager, sport a rather worryingly thick moustache.

The director in the control booth feels a little warm leakage of pleasure in her knickers at the spectacle before her.

'Vor-tex! Vor-tex! Vor-tex!' they all chant as the winner takes a bow.

Then a new chant begins: 'DIE! DIE! DIE!' It is accompanied by the stomping of feet and the regular rhythmic clap-clap-clap of hundreds of pairs of hands.

Black-clad figures with guns now stand around the ring, only the safety ropes between them and their quarry. Suddenly, the safety barrier falls - collapses into the stage - and there they are, face to face, Red Devil and her killers.

'KILL! KILL! KILL!' scream the audience.

There is nothing Red Devil can do now, so why fight?

Then, suddenly, the figures start shooting. The loud stark chatter of rapid-action gunfire *tat-tat-tat-tat-tats* in the air - the familiar sound of war films and news reports, right there in the arena, being filmed and broadcast live by the BBC.

But the sound is not coming from the guns at all but from the sound system. And the guns themselves may well be shooting at their victim, but there are no bullets flying through the air, because the guns are not firing any bullets at all.

Instead, long jets of foam and gunge of all colours - bright pinks, greens, oranges, yellows - shoots through the air and covers Red Devil in gloop. She slips over, tries to get up, then slips over again.

Before long the figures in black are also covered in gunge and are slipping over too. The audience roars with laughter as bright lights flash and upbeat comedy music plays on the sound system. The camera shifts between shots of audience members cracking up and the comic sight of a glorified custard-pie fight on-stage.

Sunday, Sunday

'Another winner, Minty. Yay!' says Toby, clapping.

'Incredibly good,' says Oliver Allcock.

'Fantastic!' tweets Sir Peter Baztanza.

But one thing is worrying Minty, despite the buzz of the show and the positive instant online reaction. Oliver sums it up:

'One thing I don't understand is why X-TV just showed its usual shows. Incredibly odd, really.'

Why? Why didn't X-TV go head to head with the BBC's autumn schedule - which was publicised well in advance? Why hadn't they broadcast something special? It was odd and unexpected.

'They're planning something, I just fucking know it,' says Minty.

That Sunday evening, when scanning through the TV channels, she discovers what X-TV have been planning all along.

*

Experience is the child of thought, and thought is the child of action.

X-TV did not wait for permission to begin broadcasting, or for the approval of committees and governments who wish to control and limit human freedoms.

X-TV just did it: attempted the impossible, because it was there.

I, Rasmus, knew the challenge, and was prepared for it. I knew too that the BBC would do whatever it could to halt the inevitable and inexorable rise of X-TV.

But more importantly, I knew that X-TV could make programmes that would appeal to people more than any other channel – because X-TV is and has always been dedicated to giving the people what they want, when they want and how they want it. Morality is for religions, and individuals' own consciences: it is not for a TV company to decide what people can and cannot watch on their screens.

Every man's work is a portrait of himself. Thus, in X-TV I see my own image and the future which awaits us – a brave new world of unlimited individual freedom to a level undreamt of in the long, brutal and horrific history of humanity.

Attempt the impossible: it is all there is.
I am Rasmus, and this is my reality.

*

Who's My Baby?

Russia: a riddle wrapped in a mystery inside an enigma. Perhaps. A lawless land of 'anything goes'. Certainly.

A place of beautiful begging boys and gold-digging grateful girls, and of parents so desperate they'd be prepared to sell their children to the highest bidder for a better life. The perfect fit, then, for X-TV and its new show.

'Who's My Baby?' airs on Sunday evening peak-time – the traditional come-down day on TV, when the hectic game shows of Friday and Saturday fade into the pastel shades of costume drama or safe soapy series, often either set in a big country house or the Scottish Highlands, and sometimes both.

X-TV has not advertised the show in advance at all. Instead, its social network marketing machine blitzes the computers, laptops, mobiles and smart phones of millions upon millions of potential viewers at the precise moment that the broadcast starts.

Minty and Rasmus are watching this, back in London. Together alone.

In a little known concrete city with an unpronounceable name, not far from the Steppes of Central Asia, the new X-TV show is broadcast live across the globe.

'Welcome, everyone, to Who's My Baby?' says Natalia Nudikova, the former porn actress and new X-TV host for all productions filmed in the all-Russia region is co-hosting tonight's show with Alicia McVicar.

'Bring it on!' screams Alicia, to a studio audience who cheer and clap and scream.

'Yay! Cool! Bring it on!' says Natalia to the cheering crowd, mimicking as accurately as she can the American TV hosts she adores.

'Amazing! Yay!' says Alicia, frankly rather pissed off that this blonde bimbo is using her trademark catchphrase.

Their fixed rictus smiles spark a certain tense electricity, which makes for great TV: hate works just as well as love on live telly. Better, even.

'So,' says Natalia, 'you place the bids now!'

'Bring it on for Sasha!' screams Alicia.

There is the usual tick-tock music placed into the studio and the audience sits expectantly, waiting for how much members of the public out there, watching this on TV all over the world, will bid to own a four-year-old boy called Sasha who has been living in a Russian children's home since his alcoholic mother stabbed his alcoholic father to death with a kitchen knife before cutting her own throat – which, in the opinion of her surviving family, were respectively the best, and the second best, things she had ever done in her life.

Time is up; the bids are in. Sasha is older than the other babies and children they have sold that evening, so the bids are expected to be lower – couples usually want babies for adoption, not damaged goods, though the paedophile market for four-year-old boys is huge, especially in America and the Arab world.

'Remember,' says Alicia when the bids are in, reading from the autocue, 'all those bidding on babies and children are vetted by our team to ensure their new parents will be right for them.'

'Two million, seven hundred fifty-seven thousand dollars!' shrieks Natalia. Higher than they'd hoped. Not as high as the eleven million paid for the prettiest baby – blond and blue-eyes – sold that evening. Higher, though, than the dark-skinned dwarf gypsy tot who made just a few hundred thousand.

By the end of the show, some three hours later, over seventy babies and young children have been sold off to the highest bidders worldwide.

In London, Rasmus knows that the show is a massive success, and the instant ratings show that, once again, X-TV has utterly trounced the BBC.

Minty watches the show until the end, rage running through her blood, her envy engorged and grown maggot-fat feeding on the spectacle on the screen until it bursts into her brain, causing her to scream a lungful of noise at the ceiling:

'Cuuuuuuuuuunt!!!' she yells at the TV screen.

Shocked at the show she may be, but her shock is not caused by any disgust at the sight of babies for sale on television. Rather, it is the result of the realisation that there must be a spy in the BBC somewhere – someone who leaked information and let Rasmus know exactly what they were planning with 'Baby Love'.

They may have been advertising it for a couple of weeks, but a show such as the X-TV one she has just seen would take longer than that to organise, so someone must have told X-TV what they at the BBC were doing.

There must, it seems, be a mole amongst them. But who?

The thought gnaws at Minty's brain like rats, and will do so throughout a tense and troubled night, a clench of claws scratching and scraping at the inside of her skull to rip her restless dreams to slivers of sad reality.

Outrage

Outrage is the predictable media response to the selling of babies on a TV show.

Outrage that it can be allowed to happen 'in this day and age'; outrage that a 'new low' has been reached in British broadcasting; outrage that the law can be so casually and easily circumvented by basing the TV show overseas and broadcasting on the Internet.

Ofcom has received almost two thousand complaints about 'Who's My Baby?' – though it also previously received a hundred or so for the BBC's 'Baby Love' – but this is as nothing compared to the well over 30,000 complaints it received for 'Animal Crackers' from a pet-loving British public.

But what is most noticeable about the 'outrage' expressed is that it is only being expressed by a sententious media – national newspapers, especially the broadsheets, and the usual commentators who, Rasmus knows, hungrily report all popular TV culture while loftily condemning it – and not by members of the Great British public themselves.

A poll shows that most people thought that what 'Who's My Baby?' was doing was a good idea – placing babies and children, who would otherwise spend their early lives in bleak, hopeless institutions, with adoptive parents and families who could maybe, just maybe, provide them with care and food, shelter and love.

So what if people paid for their baby? Many – if not, most – people could see nothing wrong with it.

But a noisy minority disagreed.

'The thin end of the wedge,' gurns a stern bishop on the TV news.

'More evidence of decline of family life in UK,' frowns a hairy Imam.

MPs of all parties ask questions in the House, with much talk of children having nightmares and how family values are under threat, and about the risk to kids from paedophiles online.

But X-TV is prepared. It has correctly anticipated this line of attack so has prepared a full report in advance, which is now distributed to all media and government departments, showing that the safety checks they are using are much more thorough than those used by any UK government authority at present. In addition, X-TV announces that half of all revenues received from the baby-buyers will go to improve the lives of orphans all over the world.

Voices of dissent still mutter and mumble but, as both Minty and Rasmus well know, they are in a tiny minority.

Within hours of the media headlines, the matter has blown over – and many an opportunistic MP, ever-conscious of the opinions of voters, is desperate to get a spot on the radio or TV news in order to claim that programmes such as 'Who's My Baby?' are evidence of a nation 'at ease with its traumas', and to pledge their support for the wonderful new adoption system pioneered by X-TV, a great British company taking great strides using great new technology at great risk and great cost to put the great back into Britain, greatly.

Mole

'A motherfucking cunt of a mole!' spits Minty.

They know it's bad – Minty's Tourette's always steps up a gear when she's annoyed.

That Monday morning, she looks as though she's hardly slept, a dark black look glaring from the eyes under her blood red hair, giving her the look of the very devil on Earth.

'Incredibly good show,' Oliver has said earlier, praising X-TV perhaps a little too much, making both Toby and Minty glare at him and mark him as a possible traitor. But then, thinks Minty, he is the Controller of BBC1, so why would he want to scupper his own channel and its prime-time shows?

And besides, he's right. 'Who's My Baby?' was, Minty knows, great TV. It had everything: sadness, happiness, jeopardy, money, love and babies. What she hates is that the BBC could never get away with filming a similar show, their being in the unfortunate position of having to obey the law.

'Incredibly gutted that they hammered us in the ratings though,' Oliver says. 'It should've been a BBC success, not an X-TV one.'

No, it's not Oliver who's the spy. Could be Sangeeta or Lucinda though, each of whom have spent most of the morning looking at their shoes or fiddling with their mobiles.

In fact, right now, as Minty looks at Sangeeta, she can see

tiny droplets of sweat sparkling on the clear brown skin of the little frown on her forehead, glistening like the flesh of an orange freshly split open.

Lucinda always looks worried and is forever fingering her smartphone in her bag, no doubt attempting to juggle the debts which, Minty secretly knows through her informers, are soaring to new bankruptcy-brushing heights.

'Fantastic shows, both of 'em,' says Peter Baztanza's grin on the video-conferencing screen. 'Gotta hand it to 'em.'

'No, I fucking well do not have to hand them anything, especially not a BBC programme idea.'

'My programme idea, actually, Minty,' says Sir Peter.

'Which you developed for me and the BBC.'

'Of course. Fantastic, wasn't it!'

'Incredible problem, the old copyright,' says Oliver.

'What?' Minty is reading through the ratings, getting ever more irritated and envious at the extent of X-TV's dominance.

'You can't copyright an idea, as we all know.'

Oliver Allcock is right. You may be able to copyright a book or a song, or perhaps even an exact TV show format. But you cannot, no matter how hard you try, copyright an idea.

'Just as well really,' he continues, 'considering how many really incredible BBC shows started with...errr...*influences*... from somewhere else.'

Sangeeta and Lucinda look up at Minty, who frowns at Baztanza's grin on the screen.

'It's not the cunting copyright I'm talking about, nor the format, it's the fact that someone...' – and as she says this her eyes scan the room, like a snake seeking out a rat – 'told Rasmus about our fucking programmes well in advance.'

Sangeeta and Lucinda look at the floor again. Toby and Oliver stare unblinking at each other like cats on a wall. Peter Baztanza grins in pixels on the video-conferencing screen.

'Cunts,' says Minty, and the others in the room wrongly assume she is talking about X-TV.

'Yeah cool wicked!' says the heavily moisturised black baby-face on the giant screen to the cheers and screams of a vast audience.

The voice booms like dance music into the Shanghai sky.

'I is da one... an' da only... Danny Mambo!'

Cheers cheers cheers. The familiar lunatic noise of a television audience driven nuts by the possibility that their faces might just appear in close-up on TV.

'I am da best so forget da rest! Coz tonight, we's gonna hit you wiv a kickin show, right here on X-TV!'

Danny Mambo stands on the stage in the centre of a huge stadium.

'And dat show is called To Die For! Dis is for real, d'ya get me?!!!'

More noisy yelling and hollering as the lights spin and the loud manic music booms its bass into the stadium.

The specially-selected audience is international and multi-coloured - white, black, Chinese - because the international broadcasters and advertisers would not be happy with all-Chinese-looking spectators and contestants. Thus a more diverse audience has been specially shipped in and accommodated in the former 2008 Olympics athletes' village.

Next to Danny stands Nancy Ng, Hong Kong star of many a Hollywood-HK-Chinese co-produced movie, who is prepared to accept life under any political system so long as she can keep her celebrity lifestyle. Fluent in Mandarin and English, she attempts to translate everything Danny says immediately afterwards, though is careful to also follow the government-approved script to the letter.

'OK, guys, cool, way to go!' says Nancy Ng with her ever-sparkling American smile, after reading out the required Mandarin translations. 'This is like, so *awesome*!'

Nancy's smile meets Danny's as they stand face to face like young lovers on the stage in front of thousands in the stadium, with millions watching live on the internet all over the world.

'To Die For' is what the show is called in The West, but the Mandarin title translates literally as 'Death to Enemies!' In this way, the harmony and happiness of society will be upheld by the entertainments on tonight's show, which is seen as an ethical and healthy influence by the Chinese government.

The show will run every day this week and climax in the grand final on Friday 8th September – the date chosen because the figure eight is lucky in China.

After a tense build-up and profiles of the contestants, which have got both the stadium audience and those watching at home primed for the action to come, the real show starts.

Danny and Nancy leave the stage which, using some complex mechanics, transforms into an arena, closely resembling those of Ancient Rome, but with artificial sand of various colours, only one of which is yellow, and which can, through digital magic, be used to display announcements, logos and advertisements. Around the ring, the hoardings are also digital screens, over which adverts and slogans in Chinese characters and English letters crawl like centipedes.

Well over a hundred cameras are supported by a balletic wire structure above the stadium and they, together with the more traditional cameras around the arena, will capture every single shot for the audience, the internet and posterity: the whole production team know they are part of history.

Gary and Anita watch the scene excitedly from the control booth, feeling that wonderful sense of frantic, elated anticipation, doom and foreboding that only live TV can produce. Debs, the third producer, has retired to the hotel, feeling unwell: apparently some local delicacy didn't agree with her at dinner the previous evening. She will spend the evening watching 'To Die For' on the hotel bedroom TV, sobbing and homesick, thinking of babies and how it was she, not Ravi or Mercy, who should have been the producer of 'Who's My Baby?'

Two men enter the ring from doors that slide open at opposite sides of the arena. Each is dressed in shiny plastic and chrome, one in red and the other in black, an eye-mask stretched across each face, gladiators for the digital age. Their arms are bare, glistening and taut with sinews straining. And each grasps a weapon – the glinting sword that will be their saviour or downfall. Then, after a countdown on the huge screens, the contest begins accompanied by the same pounding, pulsing, throbbing music that has introduced the show.

In the UK, Minty watches TV at home with Toby perched on the sofa beside her.

Oliver, Sangeeta and Lucinda also watch, together with over a quarter of the British population, which soon rises to a third when word gets round online about what X-TV are showing tonight.

Minty screams at the screen, getting as caught up in the sheer excitement of the show as anyone else.

She knows this is great and ground-breaking telly – but it is obviously based on 'Gladiator' which Peter Baztanza developed for the sole use of The BBC, though even Minty has to admit the format was originally nicked from Ancient Rome. She also knows that X-TV will be able to go much further than the BBC, UK-based and constrained by rules and regulations, ever could.

After a vigorous and thrilling battle in the arena, the victor is clear. At one side of the ring the red gladiator, whose features are Chinese, is standing over the Caucasian figure dressed in black. The Chinese in the crowd are going wild – for them, the contest is all about national pride, and they will always support the Chinese fighters, even the convicted criminals, over the rest.

The red Chinese gladiator looks up at the crowd, the streak of a silver sword in his hand, awaiting orders. The black gladiator's sword is kicked out of his reach, and he lies bleeding from a head wound on the sand, breathing rapidly, wondering how on earth he – a lad from County Durham called Lee, whose mum was a dinner lady, and who wanted nothing more than to play footie and be the next captain of England – could end up lying on his back in a giant sandpit in China awaiting his fate before a TV audience of millions. All around him, on the multicoloured sand, various adverts – for drinks, cars, fast food and high-tech gadgets – blink and flash and shimmer.

The crowd is chanting: 'Kill! Kill! Kill!' in English – no matter what their first language. The chant will instantly become a favourite in school playgrounds all over the world.

'Yo yo yo!' yells Danny Mambo, arms waving at the crowd. 'Well wicked, d'ya get me?!'

'Yay, so coo-ool. Like just awesome!' says Nancy Ng's smile, before it twists and gurns into a multi-tonal Mandarin translation.

'So now, it's dat time again, innit?!'

'Hey, it sure is, so now... for the all important vote.'

'The voting starts...NOW!!!'

Danny puts his arm around Nancy Ng. She flinches in shock – she has never been touched by a black man before, and knows her family will be shocked at the multi-racial sight.

Almost before they all realise, the minute is up, and the ticky-tocky music stops.

'OK, cool,' says Nancy, 'all the votes are in.'

In the arena, Lee, exhausted, injured, bleeding, closes his eyes and thinks of his mum and prays to Baby Jesus for the first time since primary school (year three).

Then, predictably, the word punches itself into the screen like a fist slamming into a face:

'Die!'

So, in one swift motion, Lee's heart is pierced by the tip of a sword forced into his chest by a convicted gangster from Xinjiang province, who, through happy coincidence, is also called Li.

The last thing Lee sees, as his bloody heart beats and squirts its last, is Li's narrow eyes squinting and laughing at him in the sheer enjoyment of the kill. The last ever thought that trickles with the blood out of Lee's body is the thought of ordering sweet and sour pork and special fried rice, which is the take-away his mum used to get on Saturday nights when he was a kid.

Lee, who always knew he'd be famous one day, is proved right and becomes the first person ever to be killed in gladiatorial combat on live TV - something which, when his mum has got over the shock of seeing her son killed on telly, makes her dead proud.

At least he didn't die in vain - not now he's famous.

Back in Hammersmith, Minty and Toby finish another bottle of wine.

And die of nothing but a rage to live

Rasmus finishes the online meeting with China. Everyone is delighted with the way the show has gone, though the Chinese have expressed reservations about the choice of presenter.

They have, however, very much enjoyed watching various undesirables and criminals being hacked to pieces, which can only help to build a more happy and harmonious society.

It is good that they cannot see the smirk on Thursday's face as they say this. He remembers when they used to do a lot of hacking people to pieces in Africa, though with crude machetes not silvery swords, and the foreign media used to call that 'atrocities' or 'war crimes' when they did it. But now the Chinese are doing it, they'll say it's become a force for moral good. But then, the Chinese are the new masters and hold the purse strings of the world, so can do and say - and demand - what the hell they like, from anyone, anywhere, anytime and in any way whatsoever.

'And die of nothing but a rage to live,' says Rasmus, watching another gladiator being dispatched on the muted TV screen in his office. With no sound as a distraction, the two fighters on-screen look like dancers or even lovers embroiled in some passionate game – until, that is, the public vote and the inevitable spurt of blood which extinguishes the light in their eyes forever.

He motions to Thursday to turn off the television, and looks up at the Gillray prints laughing down from the walls: they have seen it all before, those grotesques, know how base and animal our instincts always will be, whichever façade we spread thick like make-up on our true human face.

London seems quieter than usual as they drive through the traffic that evening, because everyone is home watching TV. So high are the ratings for 'To Die For' that Rasmus wouldn't be surprised if the crime rate showed a distinctive dip during its times of broadcast. Such is the power of television. He hopes it will stay like this for his birthday – he has a party planned.

As Toby and Minty watch the ever-more gruesome games from China, they can only admire the undeniable entertainment value of what they see. That TV can be addictive is hardly news; that TV showing men – and occasionally women – being chopped to pieces and disembowelled in gory close-up in a gladiatorial arena could be so much fun to watch is a genuine revelation to many watching.

If people had not experienced for themselves the joy of watching a fight to the death – the sheer real exhilaration and release of it – then they would have continued to mistakenly think of their ancestors as bloodthirsty brutes and butchers.

Now they know that we are merely the mirror image of every horror of history – slicker and more polished perhaps, and digital of course, with high definition images and surround sound. But no more different from our feral forebears than a tree in the park is different from one standing in some ancient forest from which heretics were hanged and which was felled to provide the wood for witch-burning bonfires.

*

Have human beings become monsters, a race of seemingly irredeemable creatures, hungry for blood and death?

Has television done this – changed human instinct so radically?

Of course not: the instinct is there already. The need – the yearning desire – to attack, to hurt, to kill, or to cheer on those who do, has always been there, an integral component of the human machine.

Every age has its victims and its heroes. The only thing that changes is who they are.

Show me a hero and I'll show you a tragedy.

I am Rasmus, and this is my reality.

*

To Kill For

It has been raining in Wales, and in the village of Llandoss Hugo's cottage is leaking like an old boat.

It is thus that Hugo finds himself interrupting his evening's viewing of X-TV's gladiatorial combat on his brand new digital television to go and rescue what he can from upstairs, where water is dripping from the ceiling of every room. Since the move, he hasn't yet got round to sorting out all the boxes filled with his possessions, though he suspects that most of their contents would be better on the tip anyway. So he stands on the landing hurling the sellotaped boxes down the stairs, only occasionally wincing at an occasional tell-tale crunch of china or glass.

Occasionally a box bursts its contents at the foot of the stairs – files, photos, letters, books and trinkets spilling like guts over the carpet, the detritus of a life once lived. Feeling the heavy-headed dizziness of almost a whole bottle of cheap whisky, an exhausted Hugo sits at the bottom of the stairs and starts picking up the objects on the floor before him. The photos of happier times – Hugo, Felicity and the girls in Tuscany, or France, or on one of the birthdays and Christmases when a camera inevitably appears – all smiling, all lying. Since the divorce, he has not heard anything from the children. His solicitor also mentioned the possibility of an arrest for harassment and a possible restraining order if he ever tries to contact them or his wife, or even sends them more than a couple of emails or texts – (and apparently, the police are keen on using such 'non-crime' social media laws to boost their arrest stats in these days of falling crime).

Hugo picks up the two recently-separated halves of an engraved rose bowl from the hallway floor – it's his BBC producer of the year award 1988, a time before the world went mad (though it was, perhaps, well on the way). He used to be good at it all, he knows – very good. And then...

He flings one piece, then the other, at the wall. Small diamonds of glass tinkle and ping against the brickwork, shattering with a brittle music.

Detritus spread over the floor – all the useless broken things. Then he finds the old school photos. And there he is – Hugo – aged seven or eight maybe. Thin, blond hair, shy dimpled smile, proudly-picked permanent scab on his schoolboy chin.

Hugo blinks at the faded photograph. How this boy in the photo seems so happy, so different, so alien to the man now sitting drunk and despondent at the foot of the stairs – almost as though they bore no relation to each other. How on earth had the Hugo of then grown and changed and become the Hugo of now? Just how did it happen?

There are tears in Hugo's eyes. Then he finds his old school books – the exercise books from when he was eight years old. Hugo opens one of the books at random and reads.

Nonsense poetry, mostly. But how much more nonsense there was in the real world than in that absurdity! Not bad poems at all. Eight years old and writing that – so free, so creative, so happy in the freewheeling nonsense of it all. Just where did all that go? All downhill since, really.

He continues reading:

Winter Days
The frost is cold.
The days are dark.
Here comes Jack frost in his sledge
With a little hat on his head.
People have cold ears because of him.
He goes on the buildings and pricks your ears,
And makes your eyes fill up with tears.

Under that poem, the teacher has written in red pen, 'Super work!'

Hugo closes the exercise book. His eyes now are so filled with tears that the world around him appears even more blurry than usual.

He turns his head, almost in slow motion, towards the booming sound of the television in the living room. The usual noisy noise. But Hugo's ears are humming with the rumble of nothing.

And then something strange happens: Hugo has what can only be described as a revelation.

'That's, um, it! Yes, um, that's it!' he yells, bouncing to his feet like a spring. He is so distracted and drunk that he does not feel a thing when a splinter of glass from the now-shattered BBC producer of the year rose bowl award pierces his sock and embeds itself in the sole of his left foot.

Hugo knows it. Hugo feels it. Hugo has seen the light!

'How beautiful is the world!' sings Hugo's mind to his befuddled brain, as he skips and dances around the hall. 'How wonderful is nature! How far have I wandered from my true path!'

And, at that precise moment, Hugo makes a decision: he will become a poet. For that is his true talent and always has been - since aged eight, at least. Besides, Wales is the land of poets, so it makes perfect sense really, especially with whisky still wet and wonderful in his veins.

Maybe life wasn't all futility, hopelessness and despair,

after all? Maybe life could be happy and lovely and bursting full of future possibility. Maybe he could – we all could – manage to defeat and destroy the darkness of the world and light a candle for goodness and rightness and love!

Maybe, just maybe, life could indeed be beautiful and good and true and right and wonderful – despite all the badness and unfairness and nastiness and suffering and sad, bad things which sometimes seem to fill the world like worms a corpse.

Hugo has seen the light at last!

Bursting with poetic joy in his soul, and the best part of a bottle of scotch in his stomach, Hugo flings open the front door and, standing unsteadily in the doorway, inhales the wet Welsh air deep into his lungs.

It is not raining any more – there is nearly no noise, as though God has turned down the volume, pressed a button or flicked a switch. As if to agree, a single narrow shaft of yellow light projects itself onto Suicide Mountain through the cloudy greyness of the sky. This, Hugo knows, is a sign. This, he knows, is the beginning of his new life.

Hugo runs out into the street, eyes filled with happy tears, his blurry vision blurrier for the booze – and promptly collides on the puddled pavement with a little red mobility scooter travelling at its full speed of five and a half miles per hour.

Hugo is knocked back onto the pavement, a helpless puppet whose strings have been snipped by an omnipotent hand. He watches from there as, in almost cinematic slow motion, the mobility scooter topples over flat onto its side like a felled tree.

Its occupant is Ivor Fireplace, 112-year-old former miner and dedicated cigarette smoker since the age of fifteen, and – at least until the very moment he finally stops breathing as he sits seat-belted into his mobility scooter – the oldest man in Britain (a title which now automatically passes to a Scotsman the nurses call 'Mad Malcolm', an old pervert who has spent more than forty years inside for unspeakable crimes, but who is now comfortably resident and randy in a care home in Bexhill-on-Sea).

The flummoxed look frozen on Ivor Fireplace's face when he expires is so expressive as to be almost poetic. His last thought is of a final dying cigarette burning itself to nothing in the darkness of his soul.

The Invite

Peter Baztanza paces back and forth in the Director General's office, sleek as a caged panther, his grin fixed and fearless on his face.

'Fantastic ideas,' he says. 'Fantastic! Though nothing's ever original really, as you well know Minty.'

Minty gives Baztanza a hard stare, but knows this is all too true. Like most – if not all – TV producers, she has built a good career off the back of other people's uncredited ideas.

'But how the fuck did X-TV get to know about our programme ideas?'

'Public knowledge, Minty,' smirks Baztanza. 'Word gets around. No secrets in the internet age. I'm so loving it!'

'Well, I'm so *not* fuckin' lovin' it, Peter, not with what I'm paying you.'

'Oh I thought the great British public was paying me, Minty?' says Baztanza, almost skipping back and forth in glee. 'It's they who pay for the BBC, no?'

Minty's face creases into a scowl that could crush rocks.

'They must've moved quickly,' says Toby, who stands next to Minty, a monkey loyal to his master. 'I mean...if they only found out about the BBC shows from the trailers and publicity.'

'Quick quick quick 24/7,' says Peter Baztanza. 'That's what the digital world is all about. And I'm love-love-*loving* it!'

Minty knows that X-TV's great advantage – or one of them – is the speed with which such a lithe and lean organisation can react to events, can change and adapt – and thus survive and prosper.

The BBC trailers have been airing for weeks, so X-TV could easily have copied the idea from there. And what was to stop any single member of any crew of any show letting X-TV know, through carelessness or design, what the BBC was planning. There were of course clauses in all staff contracts – but Minty knows that the only way to stop an idea leaking out is to have a workforce so absurdly well-paid that they would never ever betray their employer, and the BBC could only ever pay TV crews industry minimum.

If only everyone were like Toby, she thinks, so malleable, so loyal, so obsequious, as easy to bully and dominate as a baby.

'Make Me a Star! will be fantastic, Minty. Just fantastic!'

'Make Me a Star!' is the new BBC show starting on Saturday night. It is the third and final idea from Peter Baztanza, and is nothing more really than yet another shiny-floor TV talent show – but cruder, ruder and even more cruel. Nifty legal manoeuvres also mean the age limit has been lowered, which Minty knows will mean lots of angelic children inflicting their mangled versions of 'Bright Eyes' or 'Pie Jesu' on the nation.

'See you at the party maybe, Minty?' says Peter Baztanza, following his bright beaming grin out of the room.

Toby nods a nervous smile at his mistress.

'Oh Minty, I'm so looking forward to it...'

Toby's word trail off and he bites his lip with his frighteningly super-whitened teeth.

'Me too, Toby, me fucking too,' says Minty, a nasty smirk spreading over her face like a rash. 'In fact, I can't fucking wait.'

*

The invitations have arrived.

'Thank you, Kerry,' says Minty, too weary to be abusive that early in the morning.

'It's Kayleigh,' says Kayleigh.

'Kayleigh, Kerry, Kirsty, Curly, Crappy, Cunty…'

She opens the invitation as Kayleigh retreats from the room in silence.

It's a hand-delivered invite - expensive quality envelope, for sure. But then, as DG of the BBC Minty's always getting sycophantic missives from the tedious and the corrupt - businesses, politicians, charities, the lot - all wanting warm relations with the BBC, all sneakily trying to butter her up so they can get preferential treatment. Either that, or it's yet more mail from so-called friends she doesn't like - and she already has enough of those.

Minty carefully opens the envelope and eases out the card inside. It reads:

> *'You are cordially invited to a party to celebrate the international success of X-TV and the 33rd birthday of our founder, Rasmus Karn, on Thursday 7th September.'*

Only Rasmus would dare hold a party on a Thursday, rather than a weekend. But he knows he has the power to make people want so much to attend this party that they are willing to forego any early Friday appointments and miss work that day. Minty wonders if any BBC party would have the same effect, and is momentarily angry at having such a ridiculous thought in the first place.

The party will take place at Clevedon House, a massive mansion nestling in the lush landscaped countryside of Berkshire, west of London. Through the idle centuries, it belonged to various wealthy aristocrats and was used as a place for lavish entertaining, until becoming a hotel in the late 20th century owned by oil-rich Arabs, before finally being purchased by X-TV as part of its property portfolio.

'Incredibly beautiful gaff,' says Oliver Allcock, in the meeting later that day.

'So that's a yes then, is it?' says Minty.

'Wouldn't miss it for the world, Araminta. Incredible really what this Rasmus Karn has done in so short a time. Should be an incredible party!'

Lucinda and Sangeeta are also at the meeting, waiting to see which way to jump as always.

They think Minty is about to explode, but then her tense frown relaxes and she leans back in her chair. She considers the anxious faces of Lucinda and Sangeeta who, she knows, are desperate to dress up in their finery and go to the X-TV party at Clevedon House. It is a do which, they all know, will be attended by a great many famous people from the worlds of TV, music and film. Minty decides to allow her staff the chance to play at being celebrities for an evening.

'OK,' she says. 'We can all go to the ball, Cinders.'

Lucinda and Sangeeta - together with Toby - almost squeal in delight, they're so excited. They had thought that Minty would ban BBC employees from accepting an invitation from their main (and most hated) competitor which could only embarrass the Corporation really, and rub its nose in X-TV's stratospheric success. But Minty has her own reasons. After all, where better to see Rasmus Karn than at his own birthday party?

'It'll be a fuckload of fun, so why not?' says Minty.

After the meeting, which reaches no conclusion other than the familiar one that X-TV is hammering the BBC in the ratings, Minty asks Lucinda to stay behind. Lucinda looks worried. Her credit card debts are now huge - so huge, in fact, that she is completely ignoring them and all letters, emails and text messages that might possibly be from a credit card company or debt collector. She knows that her position is untenable - she is, after all, the BBC Head of Business Strategy. She is supposed to be good - expert, in fact - at managing money.

'If I could have a little word,' says Minty, sitting down at her desk again.

How on earth, thinks Lucinda, did she let her shopping

habits take over her life? How did she get in this huge financial mess? And how the hell is she going to get out of it?

Lucinda hasn't got a clue. Luckily, Minty has the answer.

'Close the door on your way out, Toby,' Minty says, and he does so, leaving Lucinda with the Director General, alone.

Apache

In Llandoss police station, Hugo apologises.

'I'm, um, really, um, most terribly, um, sorry, officer...'

The policeman, a gone-to-seed rugby player with a round pink face the colour of cheap processed ham twitches his nostrils at the whisky on Hugo's breath.

'Right,' says the sergeant, 'you can go 'ome now.'

'So, um, you're not going to, um, charge me?'

'I didn't say that, Mr Seymour-Smiles,' says Sergeant Gareth Glascock-Jones, wondering why these English nancy boys always had such ridiculous names. 'You has, to all *intensive* purposes, killed a man through your actions - and none other than the world famous Ivor Fireplace, may the old boy rest in peace.'

World famous? Hugo stares blankly back; no-one has mentioned this before.

'You sees, Ivor is...*was*...the oldest man in Britain at a hundred and twelve year old – a national hero in Wales, he was, and you's just finished 'im off. There could be consequences.'

Worry worms away inside Hugo's guts.

As he leaves the police station, Hugo is glad to see that at least it is not raining any more.

Then, suddenly, a camera flashes into the night. This will be the shot of the startled, shifty-looking, scruffily-bearded alcoholic that will be printed in the tabloids the next day, right next to the slick official portrait taken when he was BBC Head of Vision – before and after shots of how the mighty have fallen.

Startled by the flash from a state of partial to total blindness, Hugo lets out a little yelp of concern. There are words – talking – questions – from somewhere behind the bright white flash – asking him what happened, why, how, who he was, why he killed old Ivor Fireplace, a Welsh national hero. The usual nosey noise of hacks who smell a story – hyenas scenting the blood of the struggling.

Suddenly a hand grasps his arm and starts leading him along the pavement.

'Is you English then?' says Colorado Colwyn, when they are well away from the pack of photographers.

'Yes, um, no, um, I was...'

'So you's local now, like?'

'Yes, um, I suppose so. I live...'

'I knows where you lives.'

Hugo isn't sure he likes this. He is being forcefully marched along by goodness knows who – a man with an accent that can only be described as half heavy Welsh Valleys and half John Wayne western.

'I's Indian,' says Colorado Colwyn.

Hugo, whose sight has only partially returned as the camera-flash-burn on his retina has faded, looks at the man beside him – a man whose skin looks about as ethnic and swarthy as a Nantgarw porcelain plate.

'Not *Indian* Indian, mind – not like them Pakis down by yerr at the curry house. No, I means *real* Indian – *Red* Indian – but thass racist that is, so now they says Native American to be polite, probably coz of all the arrows, like.'

Only now does Hugo notice that the man is wearing a Wild-West-style tasselled leather jacket and has three large clipped feathers sticking up from a headband fastened round his long, straight, greasy-looking Red-Indian-style hair.

'I's Apache, I is. Running Bull. Like Sitting Bull, but runnin'. But everyone calls me Colorado Colwyn.'

'Oh, are you from, um, Colorado?' says Hugo.

'No, I's from Wales,' laughs Colwyn. 'Valleys born and bred, me – but my spirit's from Colorado, like the beetle.'

'Oh, John Lennon?' asks Hugo, confused.

'Nah, boy – the Colorado beetle. When I's at technical college in the seventies, there was this government publicity campaign about that bugger, coz it was dangerous to spuds, like – and seein' as I always wore this yellow and black jumper, I sort of gets the nick-name, see?'

'Oh,' says Hugo, thinking that it all makes about as much sense as everything else that has happened that day.

'Thanks for, um, rescuing me, Colorado, um, Colwyn, from the...'

'Papparasties!'

'Yes, um...the photographers...'

'They's all buggers they is. I hates em I do, what wiv all them lies they printed bout me an' the 'orses.'

Hugo wonders what that can mean, but decides not to pry, and accepts his confusion and ignorance with complete equanimity. He is relieved that the man he is with is obviously harmless, if a bit odd, though Hugo is perfectly used to dealing with weirdos and oddballs. He has spent a lifetime working in television, after all.

More than anything else, he is immensely grateful for Colorado Colwyn's help in escaping from the paparazzi, so politeness dictates only one response. When they reach the cottage, Hugo invites his new Red Indian friend in for a drink.

Arrival @ The Party

The whole place reeks of money. Old money, new money: the sweet, special, wonderful, deadly smell of money. That certain aroma, stench, odour: that unmistakeable heavy pheromone musk blowing its full and heavy scent on the balmy September breeze.

Money money money! Yeah yeah yeah!

X-TV money, and much more besides. British money, foreign money – from China, Russia, the USA, everywhere and nowhere – money that both existed and didn't, at the same time or maybe never at all, on blingy display and flaunting itself at the party that evening at Clevedon House.

The seventh day of September – Rasmus's thirty-third birthday celebrations, and also a time to celebrate X-TV's massive and well-deserved international success. It is *the* social event of the year, for anyone who is anyone – and a few who aren't. Those nobodies hoping that, one day soon, they can be somebodies too.

Minty is being driven to the party in the BBC's DG Bentley. She is dressed in an expensive designer dress chosen for her by Toby, who sits proud beside her in dinner jacket and bow tie. He knows far more about the latest women's fashions than his boss, so for such events Minty tends to defer to him on such matters. Diamond drop earrings dangle from Minty's earlobes, and the imitation diamonds of a matching necklace sparkle their twinkly bling bright into the night. Minty knows that neither the dress nor the jewellery suit her, but she also knows what is expected of a Director General. She could hardly go to such an important evening occasion in either her usual work clothes or a standard functional outfit.

Tonight would be special – a kind of media-land celebrity-studded version of a royal visit – and so she would have to put on the feminine shoes and gladrags and baubles she hated, just to look the part, no matter how much they all pinched and itched and made her bitch.

Tonight, she knows, she will get to see Rasmus in the flesh for the first time. Nervous and annoyed, with the inklings of a headache itching her scalp, and the top of her spine pinched tight with anticipation, Minty is attempting to take her mind off it all by reading through the day's newspaper headlines:

'Ex-BBC Boss Kills Oldest Man in Britain' shout the large letters of one tabloid front page.

The accompanying photo of Hugo is a classic night-time

celebrity shot, a startled scene scooped out of the darkness, a lone, large-pupil-ed figure looking lost and dazzled in a momentary blast of white electric light.

Minty is shocked at how Hugo's appearance has changed: the face that bears the thin drawn burden of a refugee, the turkey-neck flaps of skin where the double chin used to be – (a sign of sudden weight loss maybe?) – the unkempt matted hair that is much longer now, and greyer too, and seems to have a couple of foreign bodies – (twigs or dead leaves?) – tangled up in it; the straggly dirty-looking beard that looks comic enough to be fake, as though it is hanging from those pronounced emaciated cheekbones by wires – and it, too, seems to have things living in it. He could be the BBC's very own Howard Hughes, thinks Minty, but then she notices in a corner of the photo that Hugo's finger nails are not abnormally long – on the contrary, they seem bitten to the quick.

'God, he looks like a homeless,' says Toby, relishing a good celebrity bitch at the tabloids. 'He was always such a loser.'

'Stupid cunt!' snaps Minty, loudly, making the driver's listening ears twitch up like a rabbit's, but sending a shiver of masochistic delight through Toby, who does not know or care if the insult is aimed at Hugo or him or both.

Minty feels a strange sensation within her, one she is not used to feeling: sympathy. A peculiarly strong pang of it for Hugo that twangs her heart strings until they deliver a dull muffled bass thud of emotion into her brain. She has, after all, known Hugo for years and worked with him often, though they were certainly not what could be called 'friends' – not that Minty ever cared about or needed friends anyway.

Typical Hugo, thinks Minty, her eyes fixed on those of the sad old man staring back from the tabloid front page. Trust him to manage to have a collision with a mobility scooter belonging to the oldest man in Britain. What are the chances of that? For her, or Toby, and anyone else, it just simply would not have happened. But hapless Hugo just seemed to find disaster wherever he went, as if he had some kind of homing mechanism within him, some kind of self-destruct sonar, magnetically and magically guiding him towards catastrophe.

True, Minty ponders, Hugo may well be an incompetent, wet, pathetic waste of space of a man, just like so many others at the BBC. Nevertheless, she would not have wished this fate on him, especially not as he had put up with that awful wife of his for so long – one of those typical London executive wives who'd never done a real day's work in their lives and who owed their entire wealth and standard of living to dead or divorced men: fathers and husbands and exes. How Minty hated women like that – saw them for the plastic parasites they truly were, under all that pretence of independent womanhood.

And now the papers are saying that Hugo could be arrested and charged, and even face jail, if convicted of being responsible for the death of someone who, quite obviously, should have popped his clogs years before anyway. She knows Hugo could never cope with prison. He could hardly cope with the BBC canteen, for fuck's sake!

Of course, Minty had never envied what Hugo had always described as his bucolic bliss in Wales, just as she had never envied him the monied and privileged background she had never had either: she could see how such advantage often made people weak and lacking resilience, in spite of everything.

So much for family life, thinks Minty as the car slows and swings into the drive of Clevedon House, past the flash of the paparazzi, the armed police and the crowds of fans hoping to catch a glimpse of someone famous.

Minty looks out at the noisy and eager young faces, scrunched into disappointment when they see that the car does not contain a celebrity – a reality show winner, gangsta rapper or film star.

One of them – a blushing buck-toothed teenage girl with industrial-sized braces and big hungry eyes the unblinking blue of a true fanatic – manages to wriggle over the barriers and press her face up against the car window as it passes. Her face falls in disgust as she sees two really old people in the back that she does not recognise at all – she has definitely never seen their faces on TV. The girl hates herself for her bad luck at choosing a car with nobodies in the back, and seems to give up the will to struggle as police officers lead her away sobbing and wailing. To the girl and the rest of the crowd Minty Chisum is as entirely insignificant and unimportant as anybody else who isn't famous for being on TV, although some might be slightly impressed if they knew she was actually in charge of quite a lot of it.

Minty thinks of the look in the girl's eyes as the car speeds on down the drive. At that age, would Minty really have behaved like that, been utterly obsessed with celebrities, as all teenagers seemed to be now? She didn't know, but she didn't think so. But maybe if she were young now, she would be. There was just so much less TV back then, and no internet either of course, and even the newspapers didn't have a celebrity obsession. In fact, she didn't think the word 'celebrity' even existed in those days, and if it did, she'd certainly never heard it spoken.

Famous people – that's what they had back then – like film stars or pop stars. Really famous people, with talent. People who had actually done things to deserve their fame – movie stars, or pop stars and musicians who could really play and sing. Not talentless attention-seekers who were famous for being famous on some reality TV show. But that's just how the process of celebrity works these days: young, lonely, vulnerable kids with little else in their lives like to associate themselves with TV celebrities and vicariously enjoy their lives and lifestyles, to fantasise about a better and happier life without sadness, loneliness and worry which – in an elegantly intricate irony – never actually exists for the celebrities themselves.

That bucktoothed girl with the braces – who was she? Who were her parents? What was her life? Minty thinks how she could have been one of Hugo's daughters - same age, same obsessions, same complete disregard and thoughtlessness and incomprehension about how much their parents love them. Family life. How Minty hates family life - the hypocrisy of it, the hurt, the pain. She knows full well that it is that family life - *that* wife and the children they spawned together - which has been responsible for destroying Hugo.

Or maybe it was something else that did that. Maybe television itself was the brutal and all-consuming ever-ravenous beast that killed the beauty of life for an innocent, inept and pathetic - but basically good - man.

Or maybe, just maybe, Hugo had just been a stupid bloody idiot to get himself in such a mess.

The Golden Bull

Thursday stands in the shadows, watching.

This has to be done, he knows, as does Rasmus. The private detective hired by Minty Chisum had been getting far too close lately.

Minty has no secrets from Rasmus. Whether she knows it or not, she lets them escape from her brain into an insecure world. All her texts, emails, phone calls - everything on any database digital system anywhere, as well as many conversations - it has all been accessed and analysed by X-TV. So Rasmus knew about the private detective who had been sent out to investigate him and his identity almost as soon as Minty hired him. It was a pointless move by Minty, and it was bound to end badly.

A man steps out of a pub, The Golden Bull, a favourite of police informers, tucked away in a leafy corner of south-east London and well away from any CCTV.

He does not know what hits him.

With one hard windmill swing of his arm, Thursday brings down the axe on the man's head. The detective falls forward, the blade stuck in his skull. Dead before he hits the dirt.

The case will always remain unsolved. Rasmus has done his research – the private detective had also been investigating police corruption and there are plenty of officers, serving and retired, who have an interest in his demise and in keeping things quiet.

Some mysteries are meant to remain so.

The Party

The car glides to a halt in the substantial parking bay to the west of Clevedon House and its neatly laid out gardens.

The whole place is set back a long way from the road, too far for any paparazzi lens to focus, especially as there are trees blocking the view on each side of the estate. CCTV cameras watch everything, and sensors monitor activity all around the outer wall. Acoustic devices block all sound too, to stop long-range microphones eavesdropping.

As late afternoon melts into dusk, floodlight bathes everything in just-right white light, the sort of soft yolky yellow glow that makes everyone feel comfortable and safe, like being on the inside of a giant egg. It means all those who want to be seen can be, and all those who don't can skulk surreptitiously in the dark corners, observing the action around them, like cats. No-one has been left blinded and squinting in the floodlit halogen glare, as would have happened at any BBC outdoor event, Minty knows.

Clearly, as much thought has gone into getting the lighting just right as has gone into everything else that evening – the food, the drink, the waiting and security staff, the music and entertainments, the whole *feel* of the thing. Because, as Minty soon realises, everything about the party is as near perfect as she has ever known it: as a TV producer, she knows effective, slick efficiency when she sees it. The party is a living work of art – a grand decadent production that has only one thing missing: television cameras. Apparently, they have been banned by the express orders of X-TV itself, together with any form of camera or recording equipment, though she assumes that X-TV are surreptitiously filming the whole thing. With so many celebrities around, no-one wants to take chances.

'Wow,' says Toby, eyes twinkling as wide as a child's on Christmas morning as he surveys the herds of celebrities grazing on canapés before him. 'Just...wow! O.M.G. – isn't that...?'

'Oh do shut the fuck up,' grumps Minty, though she knows exactly what he means.

She can see the celebrities too, and the stars – the *really* famous people – mingling amongst the many there who are not 'rich and famous' but just very *very* 'rich'.

And then there are those, like her, who are not famous, and not particularly rich, but who are involved in the production of a product – television – that exists, for most of the time, to glorify both fame and money. She can see them there, all the familiar faces from the BBC, ITV and the rest, all emerged like cockroaches from the dank and dismal dreary corners of the media: producers, directors, executives, planners, schedulers, compliance officers, journalists, PR people, press officers. Politicians even. And they are all – every single one of them – absolutely loving it, all desperately grateful to be amongst the chosen few allowed to attend such a gathering of society's celebrity elite.

There are numerous foreign faces there too – Chinese, Slavic, African, Asian, other – amongst both waiters and guests, and a babble of languages laughed through the evening air in the stilted guttural tones of accented English of varying levels of competence and proficiency.

And, inevitably, accompanying the many men there who look as though a million pounds would represent nothing but small change are the women – Russian and East European girls mostly, all with American hairstyles to match their affected LA accents. They are all dripped into their designer dresses as thin as rain.

Minty and Toby make their way towards the main marquee through the formal gardens at the front of Clevedon House. Minty notices the security straight away, though she can see they are trying to be discreet: mostly black men, and some women, dressed in dark jumpsuits or sometimes smart suits, with walkie talkies and earpieces, and the observant restless shifty eyes of those used to protecting others.

They could not be more different to the waiters, mostly young and beautiful, black and white and Oriental, the girls dressed skimpily, the boys bare-chested and beautiful, all smiling a well-rehearsed warm and subservient welcome to the guests with trays of canapés and caviar. There are also several girl and boy waiters who are sprayed gold - and they move through the crowd with trays of champagne like real-life Oscars. And then there are the dwarfs.

A dwarf dressed as a penguin offers Minty and Toby a glass of champagne as they admire a huge ice sculpture in the shape of a tiger. It is one of several arranged in the gardens, mostly of animals - eagles, lions, polar bears - though there is one giant ice sculpture of the X-TV logo standing at the far end of the garden, bathed in the light of the marquee. Minty hadn't noticed the penguins before, but now she can see there are lots of them - flocks of them - little costumed black-and-white homunculi swerving this way and that through the sea of celebrities.

But there is one person Minty cannot see, and she cannot see him because he is not there: Rasmus.

'Incredible do,' says Oliver Allcock, dickie bow skewwhiff, popping a softshell crab canapé into his mouth.

'Jussincrell...bll'.

Oliver sips his champagne and swallows, unfamiliar flavours tangy and moist in his mouth. Sangeeta and Lucinda, almost embarrassed in their expensive evening gowns and cheaper costume jewellery, smile sheepishly at Minty, their ultimate boss.

Oh how they must love it, she thinks, these young women who have not had to work half as hard as Minty to get where they are, what with all their 'positive action' and 'diversity schemes'. But Minty does not envy them. It is her struggle which has made her tough, made her strong, not weak and teary like some of the girls she sees these days.

Minty can see there is money and fame here, can see the predictable effect it is having on everyone. Is she really the only one not to be overwhelmed and impressed by it all, to think there are more important things to think about in the world than the brittle trivia of celebrity? She knows that this is a product of her upbringing, knows she will never be able to enjoy things like this, fame like this, money like this, even if she wants to – not when she was brought up to see such ostentation as a vulgar affectation to be pitied rather than admired. What is perhaps the strangest thing of all is that Minty is the one considered odd these days for believing there is more to life than such transient fame.

Out of the corner of her eye, Minty sees Calvin Snow entering the main building. She knows him from the X-TV shows. Wonders how he ever got caught up in the world of TV and wonders what he'd be doing if he hadn't become what most young people these days seem to want to be – a television presenter. Wonders if he is happy, proud of his achievements. Wonders what his mother is thinking…

'Fantastic!' says a familiar voice. 'I *am* so loving this!'

Sir Peter Baztanza is slinking his grin between the guests, sleek as a careful cat, purring at the presence of others even richer than he is, just as others purr and simper at the sight of him.

Minty can see Penelope Plunch, Head of Compliance at the BBC, squealing with laughter and blushing like a schoolgirl as some bald Russian Oligarch tells her a story or a joke – or perhaps just the telephone-number figure in his bank account. It is a rare appearance from Penelope, and proof that she actually exists to those there who have never met her.

A small softshell crab melts its saltiness onto the roof of Minty's mouth, seeping its seascape of flavours into her.

She chews and watches the world - his world - Rasmus's world - not hers. It can never really be hers, all this wealth and fame, even as she walks upon its surface. But maybe it's not his either. Rasmus is another one not interested in all this superficial vacuous showiness - all these baubles and trinkets and trash. Minty knows that as well as she knows herself.

Oh to know what is beating in that unpoetic heart of his!

Rasmus. Where the hell is he? Minty looks around, scans every single floodlit face. Definitely not here. Not down here, with the guests, anyway. But she knows he's here somewhere. She can feel him watching - feel the heat of her image burning onto his retina at that very moment.

'Plenty of *dough-ray-me* on display with the squillionaires tonight, eh, Minty?' says Oliver, as if reading her thoughts. 'Can't see the birthday boy though - seems to be missing his own party. Incredible eh?'

'Who the fuck does he think he is - Gatsby?' says Minty, waving away another penguin in the manner of a surly sea lion.

There are only so many glasses of champagne, softshell crab and caviar canapés that she can take. 'Don't these midgets have any dignity?' she thinks, before glimpsing recognition of one of the waiter's faces from the hit X-TV show 'Dwarf Orgy'.

Just then, a pounding drumbeat starts: the band has started to play. Its loud Indie-rock displaces the gentle pretty pop that has been piped into the party as the guests have arrived, a soundtrack designed to reassure and caress like foreplay.

But now *The Poisonous Little Pumpkins* are on stage - a group initially famous for being the most shameless drunken drug-addled shambles of an Indie rock band ever, before cleaning up their act and becoming the go-to guys for a great live performances of both their back catalogue, and just about every and any great pop/rock song ever written.

The manic though woefully tuneless lead singer, 'Dopey' Denzil Pencil, bops his mosh like a mentalist, whilst the dark and mercurial Kirk Kinetic, guitar hero of a generation, bleeds every gothic lick he can out of his loud death-black electric guitar, weaving its melody between a solid wall of bass courtesy of bass-king, Pete Van Slam. The pounding snare drum snaps into 'Walking on Sunshine', guitars brash and bright as the gleeful cheer emanating from the singalong crowd, and a bassline to die for.

Several of the Russian oligarchs' girls ask their permission to go closer to the stage. A nod, a kiss, a girly piglet squeal and off they go, tottering on their heels, fake tits jiggling their silicone jelly, stick-thin arms waving in the air like reeds as they start dancing forwards towards the stage. It is just like the old days in the clubs of Moscow and St Petersburg, except this time they have clothes on and they actually know the name of the man who will fuck them that night.

Fortunately, the music is not loud enough to interrupt the conversations of those who do not wish to go to a rock concert, and the acoustic design so advanced that those on one side of the gardens can hear the live music loud and clear whilst those a mere few feet away, just past the middle of the gardens, hear merely a pleasant background party noise over which to chat.

Nearby, a BBC huddle is forming, a growing and noisy blob of babble. Minty knows insecurity and smugness will make the BBC guests coagulate into a big greasy globule of gunk like this, as always happens, and that this will make them cling together all night. She is glad she is nowhere near it.

There they all are, thinks Minty, the BBC bullshitters and bureaucrats – the 'Bunch of Boring Cunts' as she calls them – all the meddlers and mediocrities and malingerers who now call her 'boss'. All the users and losers and lazy thinkers, all the uncreative uninspiring managerialist box-tickers and bean-counting production executives, all the usual tedious types who make the BBC what it now is – a slow, lumbering, lolloping, plodding, plopping, fatuously failing, flailing television company that's being completely and utterly outclassed by its far more popular, successful and entertaining competitor X-TV.

There's Kevin Kureishi - over there - Head of BBC News Commissioning, chatting away to Sunday Omotayo, star news presenter and probably the blackest man on TV, whose arms wave this way and that in the air in the manner of the upper branches of a proud persimmon tree in a hurricane.

Next to them sits Suzie Colquhuquhun in her wheelchair - her arthritis was deemed only mild at one time, but suddenly and rather coincidentally got a great deal worse when, as a young and ambitious TV producer, she learnt what an advantage a disability could be to her BBC career ambitions. And she was right. She is now Head of Disability Awareness on a fat six-figure salary.

Suzie stares at the solidly substantial crotch of Jonty Bonchance de Peloux-Menage, who, she knows, got his big break at the BBC from an outreach programme designed to recruit more working class people into The Corporation. In reality, Jonty, when aged sixteen and thinking of future opportunities at the BBC, left his rather pricey public school specifically in order to enrol at the worst further education college in London. He so impressed the interviewers with his tale of hardship on 'the street' and the estate where he rented a room that he was offered a job on the spot.

Jonty chats to Poppy Popat, Perdita Heppelstein, Monty Malik, Hermione Pinchbeck and Mohammed Mahmood, all senior BBC executives just like him, about being a senior executive at the BBC, which they all find simply fascinating.

Oh, and now Judy Loris, Head of Counselling at the BBC and an annoyingly clingy woman whom everyone tries to avoid, is crouching down to ask if Suzie Colquhuquhun is alright and would she like a drink?

As Judy leaves to locate a bottle of Peruvian mineral water - (she made the drinks request difficult on purpose) - Suzie wonders how on earth this terminally unconfident and annoying woman could possibly think she could be of counselling help to anybody, what with her philandering husband, her suicidal dope-addled son and her two anorexic daughters.

Oh good, thinks Suzie, here come India and Apakshit Amorliwala, the star newsreaders of BBC1, a golden couple who dominate all primetime news programmes these days.

Graham O'Nions stands proudly next to them – (they were his discovery as Head of BBC Outreach) – and Poppy Jasperplax, who leapfrogged over very many better men to become Head of Policy and Content, is enjoying her success and that of all her colleagues, many of whom owe their careers to apparent race or gender 'disadvantage', yet most of whom were actually privately educated and from very well-off and well-connected backgrounds indeed.

Charlotte Carpark and Marjorie Meltdown, often called the sexiest newsreaders on the BBC, chat and chirrup away, trying not to show how much they hate and envy each other.

Percy Plumbago-Wang, young half-Chinese presenter of gardening shows chats happily away to his Nigerian-heritage girlfriend Andi Omo, terminally unfunny comedienne with her own TV show on BBC2.

Nearby Emily Egg, BBC sex education presenter is talking to her producer Sandra Salt, and near to them, Perseus Pepper is showing off by speaking French to Jean-Bertrand Boeuf, star presenter of BBC food shows, who speaks excellent English but who deliberately exaggerated his Gallic accent to custard-thick *sauce anglaise* proportions when he found what an advantage that could be in British broadcasting.

Deborah Dinkel, Head of BBC Education arrives on the scene, adding to the babbling blob, another microbe in the mix. Deborah owes her career success to the fact that she used to be a man called Darren until the detachment of his inconvenient dinkle during a career-boosting sex change operation led to the creation of Deborah.

Then Wilhemina Kong, Head of Regional Drama, and Paul Knowall, Head of the whole BBC Drama department, arrive arm in arm, each looking as smug as the other, like the theatrical luvvies they so wanted to be, each pretending they're at the Oscars.

They are soon regaled by the sesquipedalian Marcus Boreham-Hall, the BBC's main commentator on all matters cultural - and host of late night review, early review, mid-afternoon review, amongst others - whose face seems mostly to be made of beard, almost right up to his eyes, so much so that the little patch of blinking pink face visible resembles nothing less than some strange monkey's upside-down bottom poking through a mass of thick simian fur. Marcus is droning on, as he always does, about the 'discombobulatory juxtaposition' of something or other (film, novel, artwork), 'with the post-modern paradigm.'

Minty witnesses the growth of this huddle of hypocritical hum-drum-ity with complete and utter undisguised contempt. She turns away and sees out of the corner of her eye a group of dark shifty-looking figures standing huddled at the edge of the lawn, briefly illuminated in a cave of orange light as cigarettes are lit.

These, she knows, are the writers - the lowest of the low, in the world of TV - desperate slaves who are treated like dirt by everyone else, despite the fact that they are the ones who initially create most TV programmes. There would simply be no TV without them - no drama, no movies, no soaps - and no other programmes either, because all ideas have to be written down, scripted and developed. And yet, these creators are not respected, praised or valued. Instead, they occupy a position below anyone else in the TV hierarchy, way below the presenters and actors - (always referred to in TV-land as 'the talent') - who read the lines the unseen writers have written; way below the directors and producers who profit from their toil; below even the production assistants and runners, who at least are on proper salaries with benefits, and not permanently and hungrily desperate for whatever crumbs any commissioning editor or producer decides to throw from their fickle funding table.

Such is the lot of a writer, as Minty well knows, having put up with thousands of writers - and wannabe writers -

when BBC Head of Drama. But these are the lucky ones – the ones whose writing gets paid for and produced, though their positions as writers are so precarious that they'll seemingly tolerate any insult to stay employed and commissioned. She watches them huddle in the shadows, taking deep drags of their cigarettes, drinking themselves ever drunker and eating themselves even fatter than they are already. She knows what they'll be talking about too. Money. It is what writers talk about before anything else – always. The only thing Minty finds strange is that anyone would invite any writers to such a party in the first place – they are, in general, not known for their good looks or sparkling conversation, which usually revolves around how hard their lives are and how little they earn.

'Hey, sister,' says a voice. It is Bonnie Bojangles, African-American feminist arts commentator and writer, whose flatlining career was given a kickstart when she crossed the Atlantic and starting making BBC contacts on the arts scene.

'Y'all should be enjoying the party here, General, having you some fun.'

'*Director* General,' clarifies Minty.

'Hey, sister, my daddy's ex-military, so I understand just what you brothers and sisters gotta do, the pressure you're under, day in day out, in times good and bad, through prosperity and adversity, enduring the lash of the whip and the bitter swill of war, and whatever the Lord sees fit to send us, but...'

'Bonnie!' squeals a squeaky shrieky shrill feminine voice. Oh thank fuck for that, thinks Minty.

'Hey DeeDee, you look fabulous, sister.'

DeeDee Zeetels, Head of BBC Four, leads Bonnie away to meet Gordon Goodenough, boss of Channel 5, whom she's spotted at the far corner of the gardens. Minty sees that he's arguing with a greasy little man she's met before – called Piers or Miles or Piles or something – and also with Gillian Jizz, famous TV nutritionist, who was stripped of her 'Dr' title after it was revealed she'd bought her PhD in 'Nutritional Science' off the internet.

Toby returns blushing from where he has been chatting with Gavin Gogaye, reality TV star and ex-bricklayer, whose experience in the Big Brother house seems to have given yet more credence to theories of nominal predeterminism. He is just about to say how he has always fancied Gavin and now has his number so is going to hook up with him later, he is sure, when a clear authoritative voice interjects:

'Minty Chisum, Director General of the great BBC – who'd have thought it!'

It is Lesley Hoppity, former controller of BBC1 and a woman Minty loathes, and not only for once being in competition with her – at least on paper – as a possible future DG.

No – thinks Minty – Lesley Hoppity is nothing more than a jobbing manager, with no real understanding of the medium of television, and no real talent. She should have been managing shelf-stackers at a supermarket, not in charge of everyone at BBC1.

Hoppity's tenure at BBC1 was marked by the sacking of as many male managers and producers as possible and their replacement by relatively inexperienced women. Many industry insiders were well aware that the consequent crappy, sloppy programmes and plummeting ratings were directly caused by this policy. For contributing so effectively to the ruination of the BBC, Hoppity was given a million-pound pension package when she was eventually forced out, plus a shiny CBE from The Palace, and directorships with several City firms.

'Fuck off, Hoppity,' says Minty, not seeing the need for small talk.

Hoppity leans forward to whisper:

'I earn five times what you do these days, Minty. Just call me Dame Lesley after the New Year's Honours – there's a good frigid little bitch.'

If the hard nudge Lesley felt a moment later in her ovaries had been discussed at an employment tribunal, Minty would have sworn on the Bible, the Koran and even the Highway Code, that it was purely accidental and absolutely unintentional.

'I hate that woman,' says Toby, watching 'Greedy Lesley' limp away in pain, before remembering Gavin Gogaye. 'Minty, I've just...'

'Shut up, fuckface,' she says. 'Follow me.'

Toby wonders if Gavin will be able to make him shudder like Minty at some indeterminate time in the future, preferably when they are both naked.

'Scuse me,' says Toby, as Minty pushes past Finella Furskein, (famous artist and winner of the Turner prize for her menstrual flowers – the prettily gory patterns printed in blood by her vulva), who is now admired by the art world for her creation of invisible art, such as her artwork 'INVISIBLE' – a blank canvas she stared at every day for three years and which recently sold for three million dollars at auction in New York. Her much-emulated 'Invisible Sculptures' have been purchased by several major galleries worldwide (though apparently one was lost in transit, and another stolen to order).

'Sorry, can we...?' says Toby to Vicky Verky and Wendy Windows, the female comedy duo with their own BBC2 show, and the most side-splittingly thing in female sketch-based comedy since French and Saunders sat on armchairs talking to each other about French and Saunders.

'Can we...?' says Toby as they push past Cosmo Spitz, famous chef, and his half-Chinese wife, Charmaine – whom everyone wants to call Chow Mein but doesn't for fear of being branded casually racist.

'Sorry, mate,' says Toby, trying to sound macho and straight, as he pushes into Joe Blob, famously risqué stand-up comedian from Doncaster, and occasionally functioning alcoholic, who spills some of his pint of vodka and coke, and swears that he would kick the head in of the bugger who made him do it if he wasn't so *totally fookin' bladdered.*

'Scuse, sorry,' says Toby as they squeeze past Jim Pastie and Mandy Muffit, old-school children's TV presenters.

'Oh just fuck off back to Salford,' says Minty to the CBBC

pair as she pushes past them crossly. So worried are Jim and Mandy at this, fearing that they, the only two white presenters left at CBBC, will be displaced by new fresh ethnic faces if they do not obey management orders – that they leave the party straight away and get a taxi into town to get the next train back up to Salford as instructed.

Eventually, Minty and Toby emerge from the crowd, and the front of Clevedon House stands there before them, as is has stood for kings and queens, prime ministers and presidents, millionaires and billionaires and more. They cannot help but admire its soothing symmetry, its perfect Palladian proportions, its golden ratio of Regency elegance and style.

There are two security guards standing by the door, solid as rocks. She will go up to them, demand to see Rasmus, because she is the Director General of the BBC – because she deserves to be acknowledged – because of everything.

And then, quite suddenly, before Minty has moved a foot forward, another two guards appear at her side.

'Please follow us, Ma'am,' one says. Black. American accent. Armed.

While *The Poisonous Little Pumpkins* play 'The Whole of The Moon', Minty and Toby accompany the security guards into Clevedon House.

Rasmus watches them approach and enter on a bank of CCTV screens, just as he has been watching everyone at the party all evening long. Like God.

*

And so that day is here: the day when we shall meet at last.

Minty Chisum may well recognise me now, or maybe she will not.

You can't depend on your eyes if your imagination is out of focus, and most people can never see beyond what they want and expect to see.

I remember her though. I remember her very well indeed.
I have not lied to her – nor to anyone.
I am Rasmus. That is all. There are no lies.
I am Rasmus, and this is my reality.

*

Minty, Meet Rasmus...

Minty enters a vast, dimly-lit, first-floor room which looks out onto the gardens and the party below. A bank of CCTV screens flickers on the far wall.

Rasmus stands behind his desk, silhouetted against the glass of the wide window. A corona of a halo, a floodlit aureole, blurs golden around his head. Below it – within it – the cool green eyes gently welcome the beatific hint of a smile that spans his calm, still face.

He gestures to the chair in front of his desk. Minty, unsmiling, sits.

Thursday, the memory of a murder still stabbing at his brain, beckons Toby into the room with an outstretched hand. Toby sits, as instructed, on a chair by a side wall. He is about to speak but then Minty gives him the look he knows so well, so he says nothing. Thursday stands next to him, like some Ethiopian slave standing guard in a throne room of Ancient Egypt.

Rasmus waits until all are seated, and then sits behind the large desk. From here, he can shift his vision between the panorama playing out before him through the window and the CCTV screens on the far wall showing various shots of the guests and the grounds.

Minty realises that the windows do not allow anybody outside to see in – she remembers seeing the first floor rooms of the house in total darkness when in the garden. Special glass. Must be. Clever.

Around the room hang pictures and prints that Minty recognises: Gillray. Unmistakeable, sardonic, scathingly satirical, well-sketched Gillray. The faces of Regency London leer and sneer down at her through the silence.

She hears laughing. No, cheering. The band outside has finished another song. A heavy power chord starts a heavy rock version of 'Jumping Jack Flash'. Rasmus does not move. He says nothing. Toby is just about to say something when Minty speaks:

'We meet at last, Rasmus Karn,' she says.

It is impossible for her not to sound like a villain in a James Bond movie.

'Minty Chisum,' states Rasmus, matter-of-fact, as if identifying a species of insect, 'new Director General of the BBC. Congratulations. Welcome to my little party. I do hope you have been well looked after. Would you care for a drink?'

'No, no thank you,' says Minty, looking at Toby, who shakes his head, despite his thirst.

'Of course,' says Rasmus, smiling archly.

'Happy birthday,' says Minty. And why not?

Rasmus, up until now so cool and in control, lowers his gaze coyly and acknowledges the birthday wishes with a nod.

'The party here tonight is not solely for me but to celebrate the success of X-TV and our wonderful – yet, sadly absent – producers.'

Most of Rasmus's producers are not in the UK at all that evening, but in a land far, far away in the South Pacific where they are preparing X-TV's upcoming reality show 'The Island' – all except those on duty at the studios in West London, and of course Debbie Owen-Tudor who is in a land far, far away in her own head and currently residing in a psychiatric clinic in Switzerland. She likes the view apparently: the snow-capped mountains remind her of fresh babies' nappies, all white and clean and heavenly.

'Congratulations on your success,' says Minty, following the protocol of always paying a compliment before making a demand.

Rasmus is enjoying meeting Minty again. It's been a long time.

Minty looks at the man before her, recognising some of him, but not the rest, as though the man she once met had been melted in some magical mixer and recast to create a completely different individual. But she knows she knows him. She stares at him, drinks in his features, knowing that his name will come, knowing that the memory will soon bubble up to the surface from the deep dark depths of her brain – eventually.

What happened to that private detective in that pub carpark was most unfortunate, though it has had the positive effect of making Minty reassess her priorities. Does it really matter if Minty can't yet identify Rasmus? After all, whoever Rasmus is, what matters most is what he is doing and how he can be stopped, not whether Minty ever manages to remember a name to put to the face.

The music booms from outside, the whooping and screaming heralding a Hendrixian guitar solo, and a wider smile from Rasmus, who is watching Minty closely, though he says nothing.

Now is the time – Minty says what she has to:

'I have a proposition for X-TV,' Minty says, watching that small certain smile settles on her host's lips, those bright unblinking grey-green eyes staring through her.

'I know,' says Rasmus. 'That is why you are here.'

'Minty,' whispers Toby from where he sits by the far wall.

'Shhh!' snaps his boss, scowling. She turns to Rasmus.

'I want to make a deal,' she says.

There – said it!

Minty leans back in her chair and waits, watching her host, calm and still as a stone, eyes glinting like chips of quartz lit from within.

Rasmus nods. Then he leans forward and speaks:

'X-TV is now the most successful television company in the UK, and will soon be the biggest in the world. We have interests not only in TV, which we dominate, but also online and through social media – all interests that are growing exponentially. Why, then, should we want – or need – to make a deal with you?'

Cheering from outside as 'Jumping Jack Flash' finishes. Bowie's 'Heroes' starts with a solid fourth, the major chords rising into the night like dreams.

'Because,' says Minty, 'because we're the BBC.'

Silence. That used to be enough once, back in the day. The BBC *was* television – it was all there was, and could thus rely on deference and duty and the entitlement of eminence. But now?

'And,' continues Minty, 'because we can give you a good deal which will benefit both you, and us.'

The eyes of Rasmus twinkle and laugh. Thursday's big white grin is almost a shining spotlight from the side of the room. Even Toby bites his lip.

'And,' Minty says, 'because we know who you are.'

Silence. Then Rasmus sucks in a small gulp of air in mock-surprise at the news. He blinks a smile at Thursday, who stifles a chuckle.

'You know I fucking well know you,' says Minty.

'So you say,' says Rasmus. 'So who am I?'

Minty knows this is challenge, knows he wants to see if she has put all the pieces together, worked out who Rasmus Karn is – or was.

'You're a writer.'

'A writer?' says Rasmus. 'A writer of what?'

'Plays, radio, TV... I met you before, when you were a writer. I rejected you. I know it's you. I *know* you. I just fucking *know*.'

Rasmus smiles.

'You had another name then,' says Minty, staring hard into his grey-green eyes, trying to see through to the truth behind them.

'Another name?' says Rasmus. 'What other name?'

Minty exhales in frustration. She just can't remember his name. The face, yes, even now when it's so changed. But never the name: it just wouldn't come.

Minty feels as though she's about to explode, but then the rage in her eyes fades as she remembers - remembers the man whose work she rejected all that time ago - one of so many. She cannot recall the name - and the sleek, slim, smart body and face of the man before her are not what she remembers. Her memory conjures up an image at a BBC party, of an overweight, pasty-faced wannabe writer, shifty-eyed and neurotic, a typical loser writer type she used to deal with every day when Head of Drama - the type of hack whose ideas could be easily harvested by eager producers without any credit or payment going to the writers themselves. Easy meat; easy money.

It is definitely him - that failed writer. She sees the same eyes when she surveys the face, though it is hard to envisage how the man she remembered metamorphosed into this one. She blinks at the memory, then turns back to look at Rasmus.

Now Thursday is sniggering openly, and Rasmus's smile has widened to a grin that Peter Baztanza would be proud of - a knowing, perhaps mocking, expression of delight.

'Thursday, take our guest outside please, for a few moments,' says Rasmus, gesturing to Toby.

Minty nods permission, and Toby is led away like a child.

Now Minty and Rasmus are alone, face to face, two unusual animals, different but strangely the same, in some ways at least.

'My name is Rasmus. Rasmus Karn. But my name is irrelevant. It is X-TV that matters, not me.'

'Hardly,' says Minty. 'At least not if you're being dishonest - fraudulent even - about your identity.'

'Identity, reality, unreality - what really matters to X-TV is television, not names.'

'I want to do a deal,' says Minty, her impatience clear. 'I can offer you a share of all new BBC reality TV productions.'

'Interesting,' says Rasmus, looking curiously unsurprised at this offer. 'But why would I be interested?'

'Because it's the BBC, and we can offer you a 50% share of profits, with X-TV and...'

'Oh a merger?' says Rasmus. 'So you want my extremely successful and growing television company to share all of its productions with your extremely unsuccessful and declining television company and then take a share in the resulting operation, most of whose revenues, I dare say, will come from X-TV's contribution?'

'Well...I wouldn't put it quite like that, but...yes,' says Minty, with an honesty rare in the world of television. 'It's a good fucking deal – you'd get so much from it.'

'Really?' says Rasmus.

'Really,' says Minty. 'The BBC brand name for one thing – the kudos, the history, the international reputation.'

Rasmus looks at the Director General – her red hair, her angular features, her bitter little lipsticked mouth. At that precise moment she reminds him of a clown.

Just then, the main door opens and Thursday leads Toby into the room again, a steadying hand on his shoulder. His face is pale and shocked.

Minty notices the strange change in Toby, some transformation that has happened at the hands of Thursday.

'Goodbye, Minty,' says Rasmus. 'Enjoy the show this weekend – ours, on X-TV, that is.'

Minty stands and walks to the door. Toby, visibly trembling, follows her out after taking a last, long, hard look at Rasmus – a look of wonder and fear entwined.

*

'Cunt cunt cunt cunt cunt,' mutters Minty with every step down the grand staircase.

Toby is mumbling something too:

'Rasmus...he's...but...Minty...please...he...Rasmus...Karn... he...'

'Come on cuntface!' yells Minty into Toby's ear. 'We're going home. Find the fucking car.'

The gardenful of guests has been transformed since they

left it, the atmosphere more drunken, more stoned, more under the spell of that place. There are now more people, it seems, although maybe the crowd just looks bigger, inflated by exclusivity and alcohol.

The Poisonous Little Pumpkins are finishing playing 'Losing my Religion' as an encore. A young girl – a newly famous pop star singer Minty recognises – is dancing topless and alone in the middle of the garden, her busy eyes restless as her limbs, her exerted mind finding a new kind of balance. Minty knows, as everyone else knows, that soon they'll be watching her descent, as the crash and burn of brazen fame engulfs her in an irresistible inferno, and she joins the Amy Winehouses and the Kurt Cobains of the world as she leaves it, maybe even becoming part of the '27 club' if she can stay alive for just another two years.

Then there are the horses – four white horses at each corner of the gardens, a rider on each one – young, nearly naked, two boys and two girls. And the fire-breathers, the jugglers, and all the waiters still weaving between the guests: the golden ones and the dwarf penguins. Not to mention the ice sculptures, blue and cold as murder, hardly melted at all.

Then Toby, in a daze, bumps into Michael Jackson.

'Hi, I'm Michael,' a small voice says.

He is one of four Michael Jackson lookalikes there that evening. This one will be performing his Billie Jean dance routine later. Toby says nothing. He is too busy trying to rid his addled brain of the thoughts put there earlier – thoughts that feel like maggots hatching deep within his buzzing skull.

'Fuck off Michael Jackson, you kiddie-cock-sucking cunt!' snaps Minty, pushing him away and yanking Toby after her in the direction of the car pound like a truculent toddler.

The crowd seems to go on forever, some vast forest of fame and privilege. Minty and Toby push their way through the throng of TV presenters and reality show celebs, past the oligarchs and their girls, the footballers and the fans – past Marcus Boreham-Hall who is still prattling on about the 'discombobulatory juxtaposition' of the 'post-modern paradigm' – and eventually find themselves approaching the BBC Bentley. There are no lights in the drive approaching the car pound, and Minty trips over something on the ground.

'Cunt!' she screams. The 'thing' groans.

Nearby is another little rumpled shape, another body lying where its mind has left it. She recognises that one: it's a TV presenter, one of her staff, someone who appears every week on TV lecturing others on the importance of a healthy lifestyle.

'He's Karn...' mumbles Toby, eyes dark and distracted. 'Karn – don't you see?'

'Get in,' yells Minty at him when they reach the car, annoyed that Toby had obviously taken too many drugs of some description – that is the only way to explain his bonkers behaviour.

As the car cruises down the drive towards the inevitable paparazzi and fan frenzy at the gate, Elvis starts singing 'If I Can Dream'.

Minty glances over to the stage to take one last look at the party and sees that this Elvis Presley, as well as wearing the white rhine-stoned jumpsuit, is Asian in appearance. He is wearing a big white turban, speckled with rhine-stones, and a neat black beard that matches his dark lonesome Punjabi features.

'Nice voice,' thinks Minty, as the song soars into the night, with Toby mumbling nonsense next to her and a thousand thoughts tangling in her mind.

And then she sees the boy again – Calvin Snow. He looks drunk, swaying unsteadily at the back of the audience where the music is quieter. He is engaged in animated conversation with, if Minty is not mistaken, Lucy Lott-Owen.

When they reach the gates, Toby starts sobbing. Minty cradles his head on her breast, not out of sympathy but to stop the flashing photographers getting a 'tired and emotional' pap shot for their front pages. She covers Toby with his coat, so that he remains completely hidden from the camera flashes at the gate.

'Faster, you fuck!' says Minty to the driver as they head into London. She has work to do.

If you wants it...

'You knows what you's doing wrong?' says Colorado Colwyn, 'an' why you can't grow nothin'?'

Hugo's head shakes a 'no' while his muddled whisky-soaked mind wonders how this Red Indian knows about his crop failures at all.

'You's got bad spirits, you 'as, 'terrible bad spirits, an' until you gets rid of 'em, ain't nothin' gonna grow.'

'So, um, how...um...?'

'I bet you wants to know how you gets rid of 'em, them terrible bad spirits?'

Hugo nods. Colorado Colwyn knocks back another good glug of free whisky.

'First thing, is you a Christian?' he says.

'Well, um, maybe,' mumbles Hugo, 'y'know, C of E at school, and, um, my wedding, and Midnight Mass every year...'

'So this God you prays too – have he ever worked for you, like?'

Hugo's mind flicks back through the snapshot album of his life: his lonely childhood, the hell of boarding school, the distant and disappointed father, his broken marriage, the children who don't want to know him, his dwindling financial resources, his debilitating depression, his utter failure to become an organic gardner – or to make anything grow on his vindictive acre of Valleys land. Colwyn would not have been able to see more clearly what Hugo was thinking if his head had been made of glass.

'He haven't, have he?' says Colwyn. 'You takes it from

me, boy, them Native American gods is miles better, an' if you wants it, you gets it, like what the old Apache saying says, like.'

Hugo wonders momentarily that, if this is the case, surely most Native Americans wouldn't have been wiped out by disease or slaughtered by the white man, losing all their land and ending up living miserable poverty-stricken rootless existences on barren reservations while their culture and history got appropriated by Hollywood and a century of schoolboys playing 'Cowboys and Indians'.

'Now then,' says Colwyn, a serious frown settling above his eyes and below his three feathered headdress. 'Does you want your land to be fertile an' fruitful?'

'Well, um, yes,' says Hugo. 'I does, I mean, I do...'

'Fab'lous! Thass the first step, that is, to be honest with you. Now, for the second step, like. I needs to know what bad spirits you's got 'ere.'

'Oh,' says Hugo. 'How, um, do you...'

'An' the third step...' says Colorado Colwyn, quiet as a smile, 'is when we prays to them Apache gods what is up there in the sky, coz if we prays 'ard enough, then they just has to 'elp us. Tidy!'

Half an hour later, Hugo finds himself standing stark naked in the pitch blackness of the midnight garden with an equally nude Colorado Colwyn, doing a Red Indian dance strikingly familiar to those shown on old Westerns. On their faces are smears of Marmite and Heinz salad cream – which Colwyn had said would have to do instead of real Red Indian face paint of which, apparently, there was a terrible manufacturing shortage these days on account of most of the Red Indians being dead and/or not painting their faces like what they used to in the olden days, like. But it was absolutely essential for Colwyn and Hugo to paint their faces, he said, so the gods would see they were Apaches, especially as they wouldn't be looking much in South Wales or thereabouts.

'Hai-hum-hum-hum, Hai-hum-hum-hum,' chants Colorado

Colwyn, who, Hugo notices, is covered with tattoos of horses, including a large one of running horses on his back with a tail disappearing between his buttocks.

Round and round and round they skip on the barren earth behind Hugo's cottage, the half moon, though half-hidden by clouds, illuminating the wicked world. The smells, the nakedness, the alcoholic strangeness of it all was almost erotic.

'We's got to chant louder!' shouts Colwyn.

'Hai-hum-hum-hum, Hai-hum-hum-hum,' chants Hugo, whisky-filled and woozy, but willing to give anything a try. His tongue can taste the salty creaminess of Marmite mixed with salad cream, as some has now slowly dribbled onto his lips. It's all quite pleasant really, he thinks, boyhood memories of songs around the campfire with cubs coming back, though Marmite and salad cream were usually served separately back then.

Somewhat predictably, it starts to rain – the kind of weary heavy Welsh Valleys rain that almost knocks you down flat with its relentless wet Welsh misery.

'Oh hell's bells!' yells Colorado Colwyn, stopping his energetic dance so suddenly that his testicles bounce like clackers before settling sorely back into place. 'I knows what I's done – I's done a rain dance by mistake. We's been chanting the wrong way round, see?'

And so Colorado Colwyn starts again, this time the right way round.

'Hum-hai-hai-hai, Hum-hai-hai-hai,' they chant, long into the night, until whisky and exhaustion lead them inside to armchair-embraced sleep.

In the morning, the world of Wales is transformed. Bright golden sunlight sparkles in the rainwater on the ground making pools of liquid silver, and birdsong chirrups cheerily in the treetops in a way that would lift even the heaviest human heart. This could, indeed, be a world reborn, another Eden, a land of the gods. There were, Hugo thinks, dark shadows upon the face of the Earth. But now the sunshine which brightens the colours of everything in its radiant holy glow seems all the brighter by the contrast.

Colwyn and Hugo, now wearing underpants but with faces still smeared with the stickiness of Marmite and Heinz salad cream, stumble outside into the garden. Hugo can't put his finger on it, but something has changed - some awful presence has disappeared from his garden with the rain. Some terrorful dread darkness, some unknowable evil, was exorcised there that day.

Hugo looks different. His face is no longer as dark as death, his body not hunched and weighed down by the sky. He even seems to have grown taller, rising up to greet this new bright and happy world. Colorado Colwyn smiles an arch Apache smile of satisfaction and success.

'See,' he says, 'if you wants it, you gets it. Thass the tidy Apache way, like. Fab'lous!'

*

People always believe others to be happier than they are, and thus have a need to be happier than others. To achieve such happiness requires the acquisition of fame, and the quickest and best way to acquire this is through television. In such a way, happiness is a perpetual state of being well-deceived.

There will always be those who wish to be happier, who wish to be richer, to be more famous. Indeed, those who have tasted each are slaves to a growing hunger.

Thus there are many who will risk everything – even life itself – to have the chance to fulfil their dreams on live TV.

And so let us sail into the future, towards the island that awaits.

To unpathed waters, to undreamed shores.

To a brave new world, where everything is for the best!

I am Rasmus, and this is my reality.

*

Minty turns on her smart television and immediately switches to X-TV. Everyone does these days.

And there it is – 'The Island', X-TV's new prime time show, in full swing. It is clearly X-TV's take on other jungle-set survival shows, though not similar or derivative enough for anyone to sue for copyright infringement.

Minty smiles at the irony – she can't even remember how many times she has lifted ideas from shows on other channels, always careful to be just different enough to avoid any legal copyright complications. Everyone did it, but no-one ever admitted to it. It was TV's dirty little secret, Minty always thought.

One new twist was that the contestants in 'The Island' were celebrities, but only in as much as they were ordinary members of the public who had been winners of other reality TV shows on various channels – any fame they'd enjoyed was built on that and that alone.

These reality TV celebrities are all sitting in a jungle clearing, somewhere off the coast of Papua New Guinea, discussing that most significant, important and fascinating subject – themselves.

They have all tasted the fruits of sudden, aleatoric fame already – but they all still want more, and more, and more. Their appetite for non-anonymity is insatiable.

Each contestant has also been given a million pounds just for taking part, which is not payable to those who drop out before the end. The main prize for the ultimate winner is twenty million, with five million a piece for all runners-up – in other words, those who survive till the end.

Suddenly, a strange unkempt caveman with straggly matted hair jumps out from the jungle bushes clutching two large toads. But this is no native – this is Tub, contestant and loner, who is determined to survive and who is used to foraging and living off the land, having done so for several years in the UK, both before and after he won a TV survival show. He takes the toads over to the cooking area and proceeds to gut them.

'Just like Lord of the Flies, isn't it?' says a thoughtful young man called Sanjay.

'Loads of fahkin flies, yeah,' says Benny Bun, cockney lad and all round kushti geezer.

'Fookin twat!' says Ali, the butch female bouncer with a shaved head.

These are not, it has to be said, people whose paths would usually cross.

'Hi, contestants on *The Island!*'

The tannoy booms Alicia McVicar's voice into the jungle.

'Bring it on!'

Everyone wakes up and stands, as if to attention, as a mark of deep respect to the voice of television – their saviour and their friend.

'Hiya everyone,' says Calvin on the tannoy, trying to sound perky, though he is as red-eyed and jetlagged as the contestants.

Excited whispers amongst the girls. They all fancy Calvin – except Ali, who's more of an Alicia fan.

'We're talking to you live from the X-TV yacht,' says Alicia, 'and we have a little surprise for you.'

'In fact, two surprises, like.'

'Because, right there on the island with you, is our very own Danny Mambo!'

Screams, whoops, whistles as the theme music of 'The Island' is pumped over the tannoy and Danny bounces out of the jungle foliage into the camp. He is dressed in a grass skirt, holds some sort of spear, and has a necklace and headdress of what look like human bones and teeth. It is perhaps one of the worst examples of casual racism seen on TV since the 1970s, and it was all Danny's idea – something that makes him grin with pride as he jumps up and down, making the teeth and bones in his necklace and headdress rattle!

'Oh my god!' squeal the girls.

'Fahk me!' says Benny.

'Yo. Danny Mambo in da house!' he yells, dancing around like a witch-doctor possessed.

'Hiya Danny!' say the tannoy voices of Alicia and Calvin.

'So, without further nonsense and ado, I is gonna introduce you guys to the final Island contestant – da one...'

Danny Mambo pauses. Something squawks in the jungle. Benny Bun farts. His best island mate Dwaine giggles.

'Da only...'

The contestants – Sade, Ali, Elspeth, Angharad, Dominique, Benny, Dwaine, Tub, Mickey, Sanjay and Alberto – all listen intently.

'Princess Sarah!' yells Danny as the fast frantic theme tune booms through the jungle.

Then, as if by magic, she's there, standing in front of them – the one, the only Princess Sarah: debt-ridden, disgraced and terminally unpopular member of the royal family.

'Hiya chaps!' says Princess Sarah.

'Fookin' 'ell,' says Ali.

'Fahk me,' says Benny.

'Bloody 'eck,' says the Brummie Mickey, and Sanjay nods in agreement, open-mouthed.

'Oh my god! Oh my god! Oh my god!' gasps the gaggle of girls, flapping their hands like circus seals – all except Dominique, who swears her republican loyalties in French.

'Hiya hiya hiya,' says Princess Sarah, bouncing about as if she's about to play a game of hockey at one of England's lesser public schools – just like the one she went to, actually.

Some of the contestants bow – in Benny's case, a huge decoratively theatrical bow that he once saw some gaylord do when he was a kid and was taken to a pantomime by his mum, the slag.

'Nah guys,' says Danny Mambo, waving the carved tribal stick the producers have given him to mark his authority as presenter. 'Nah, she don't want none of dat royal ting, yeah, d'ya get me? She just wanna treated nice an easy like a normal lady, innit?'

'Golly, it's hot, isn't it? says Princess Sarah. 'Anyone fancy a nice cup of tea? I've brought some organic fair trade especially.'

Make Me a Star!

Minty picks up her remote control.

She is alone at home – exactly where she wants to be for this make or break moment, the last TV show she commissioned from Peter Baztanza.

She sips her gin and tonic, knowing this is her last chance.

Sighs.

Breathes.

Presses the button on the remote control and opens her eyes in tired wonder to her widescreen TV.

'I want this *tho* much – *tho tho tho* much – *it'sth* all I ever wanted, *it'sth* my dream, and I'll never *sthtop* dreaming my dream.'

So says an overweight, awkward and spotty seventeen-year-old girl called Stacey, who works in Burger King in Southend-on-Sea, to TV presenter Tracey Turntable.

'I want the judges to *thee* the real me, show them how *muchth* love I have *insthide...*'

As Stacey must tip the scales at something approaching twenty stone, Tracey thinks that there is undoubtedly rather a lot of room inside her for several large portions of love – and probably several sides too.

'I want it *tho* much, *Trathey*,' says Stacey, 'and I know if I really *really* believe, I can make my *dreamsth* come true.'

The public know by now how these shows work – how they exist as freak shows existed in the past, how they are nothing more than a televisual version of a medieval market square where everyone can go to laugh at the village idiots and throw rotten cabbages at the stocks.

These talent shows were never about the music – or, indeed, the talent. They are and always have been about laughing at those less fortunate – the vain, the silly and the deluded – and that is always such great fun. Pain is always more entertaining than joy – and makes for far better telly every single time.

'I wanna *thsing thsomething thspecial,*' says Stacey as the music starts.

Stacey walks the long walk to the centre of the stage. There is much muttering from the audience.

'Hi,' says Javina Jelly, the nice mumsy judge.

'Hi,' says Declan O'Dunk, the passive comedy Irish one.

'So, what's your name, darling?' says Silas Snide, main judge, and "the nasty one".

'My *name'sth Thstaythee,*' says Stacy, who stands there squeezed into the tightest of leggings and leotard, looking like a giant baby bird, featherless and pink, that is about to be pushed out of the nest and left to the mercy of carnivorous ants.

'She said Stacey,' whispers Javina to Silas.

'Oh, thanks Javina,' he says, with a deliberate diphthong (to make it rhyme rather well with *vagina).*

'It's *Javeeeeeena,* as you well know, Silas Snide,' says Javina, but she is ignored.

'OK, and where're you from?'

'*Thsouthend,*' says Stacey.

'Oh, an Essex girl?' says Silas.

'*Yeth,*' says Stacey through her snaggle-toothed smile.

'And who would you like to be as successful as?'

'Whitney *Houthston,*' says Stacey, to audience giggles.

'And what're you going to sing tonight, sweetheart?' says Declan.

'I'm gonna *thsing* I Believe I Can Fly.'

'OK, big song,' says Silas Snide, leaning back in his chair

and biting his pencil – a childhood habit that years of expensive therapy has not been able to fix.

The backing track starts.

For the next two and a half minutes, Stacey from Southend attempts to instil in the audience a belief that she does believe that she can indeed fly, as well as touch the sky, and somehow perform the zoomorphic avian feat of spreading her wings too. It is not a successful attempt.

When Stacey finishes the song with a *'Fly-eye-eye'*, having given every ounce of energy to sing it, strained every vocal cord to the max, and flexed every muscle to perform for the judges, she stands back with a satisfied smile on her lisping lips, expecting to hear cheering and applause. Instead, her smile droops as she realises what is happening: the audience is laughing. And not just quiet polite *'laughing with'* laughing either. No. This is nasty ugly *'laughing at'* laughing. They are all pointing at Stacey and roaring their heads off – huge guffaws and screams of laughter launched at her like well-aimed arrows.

Silas Snide lifts a hand and the obedient audience quietens to listen to the judgement of their master.

'I have to say that that was the worst performance I think I have *ever* seen.'

Stacey's face falls flaccid in the spotlights, her brain trying to take in what her eyes see, and her ears hear – which is all so completely different from the image she has of herself and her singing abilities.

'You may be able to fly, love,' continues Silas, feeding off the hateful hiss of energy from the audience behind him, 'but you can't sing, you can't dance, you're fat, you're ugly, what more can I say? But then, showbusiness's loss is Burger King's gain.'

'Silas, you are so rude,' says mumsy Javina. 'Just ignore him, Stacey.'

Stacey tries to smile through the unhappy tears bubbling within her. It wasn't meant to be like this. When she sings, it sounds beautiful to her own ears, and everyone says so too – well, her mum does, anyway, and at least two aunties.

'I thought you were dreadful, so I did,' says Declan O'Dunk, 'just awful. Like Whitney Houston? Maybe underwater, when she was drowning in the bath! My advice if you think you can fly is to find a very large building and jump off it...'

'Declan!'

'Then at least we wouldn't have to listen to that awful noise again!'

'Declan doesn't like you, darling,' says Silas, 'and nor do I.'

'But,' says Stacey, '*thisth isth* my dream, I want it *thso thso* much...'

'And that lisp,' says Declan, 'thank goodneth you didn't *thsing "Thsay Thsay Thsay"*, I *thsuppose.*'

The audience erupts into wicked and gleeful cackling.

'Sorry, Stacey,' says Javina, kindly, 'but it's a no from me too.'

A tearful Stacey waddles off the stage to a thousand grinning audience faces, twisted into the shape of the mob, all looking and laughing and pointing at her in a spirit of evil glee.

Fabulous

'This is fucking faaaaaaabbulous!' shouts Minty at her TV.

Peter Baztanza's idea was simple: do what all other talent shows have been doing but make it harsher – more cruel and nasty. All TV talent shows are based on the same dubious formula – wishful thinking and trampled dreams, with a sprinkling of fairytale success sitting on top representing the one in a million who will prove good enough to make a career of their talent and become genuinely rich and famous.

So 'Make Me a Star!' will follow the usual format, choose the victims to humiliate, select the cripples and blind singers, and those with "pity party" stories to tell, some of whom can be earmarked as finalists – but do it all much more nastily, spitefully and vindictively than anyone else has ever done before. And, last but certainly not least, use children.

There are strict rules determining if and when children can take part in such shows or appear on television at all, but Baztanza's plan was genius. If they made the whole performance side of the show a charity, and registered it as a school, then they could count the performances of children on the show as part of their school curriculum and even as part of their coursework for assessment in GCSEs and SATs. In this way, the rules could be wriggled around, so 'Make Me a Star!' will have entrants of all ages. The youngest to enter so far is just three years old.

Minty knows this is her last chance, and she intends to enjoy the ride. But Toby is no longer there to share it with her. After the party and what he saw or was told there, and the weird sort of mental collapse he seemed to have, he said he had decided to leave TV – in fact, to leave modern life completely – and live a contemplative life as a hermit on a Scottish Island where, as far as Minty knows, he is at this very moment, contemplating things. He left London on the morning after the party, saying he wanted to be happy.

Happy? *Happy?!*

Happiness is for the brain-damaged and the dead. It is the sole solipsistic ambition of the vain, the silly and the feeble-minded. It is thus not a fit aspiration for thinking, intelligent people who know they can never be happy nor should ever try to be, as an end in itself. Minty Chisum does not *do* happy.

Minty sees Toby's 'revelation' as nothing more than a boring breakdown of the kind that weak men are liable to suffer. She expected more of him. But then, that's what people always do, let you down. You'll never get any true loyalty from any human being. If you want loyalty, get a puppy – or better still, a pot plant.

For some reason, Toby had got it into his head that Rasmus was some sort of messiah figure. As he said on the evening he came round to explain his decision to Minty:

'Rasmus means Beloved; Karn means Carne – i.e. the flesh of God. So Rasmus Karn means beloved made flesh. Do you know what that means, Minty? Do you?'

Minty told Toby that she didn't – or rather, she did. She knew that it meant Toby had finally gone off with the fairies and was suffering some kind of psychotic delusion, probably signifying that he was developing paranoid schizophrenia and having a total mental full-on bonkers nervous breakdown.

'No, you don't understand,' Toby pleaded. 'It means Rasmus is the reincarnation of our Lord God or Allah or Jehovah. He is The One. He is the Messiah!'

'Messiah?' said Minty, rolling her cynical eyes to heaven. 'He's not the Messiah, he's a TV company owner and a total fucking cunt, okay?'

But it was no good:

'He is the One. He is the End of Days. He is All. He is Rasmus...' Toby's eyes were glazed and distant like a druggie's – or perhaps a weather girl's – glassy and remote.

Minty hated it when people went all nutty religious, though she'd seen it a few times in BBC staff, and very many times amongst needy and insecure actors, pop stars and TV presenters.

'I can see, I mean *really* see, now. There are things going on Minty, out there in the universe, that I can see, but you can't, I know that, but oh I can see it all and it's all so beautiful, Minty, so interconnected.'

'Yeah, that'll be the Internet, fuckface.'

'No, no, you don't see...I mean...Synergy...energy... interconnection...beautiful...But dangerous too...Minty,you must leave...leave London....leave the BBC...before it's too late...'

'Oh really, or what?' said Minty.

'Or when The One takes over...Rasmus Karn...the Beloved Reincarnate...bad things will happen...Really *really* bad things...'

'See what happens when you pop too many pills,' was how Minty replied.

But Toby denied taking drugs – well, apart from that E at a club a fortnight before anyway.

Minty said her goodbyes to Toby and wished him well, in her own sort of way:

'Fuck off, cunt-face. Keep in touch. With reality, I mean.'

It had been good while it lasted, but as Minty well knew, all things came to an end, good or bad.

'It's only TV, Minty,' said Toby. 'It's only TV.'

These had been his last words to her, ones which she would not have tolerated if he had still been her assistant.

It's Only TV...

Only TV? *Only TV?!* But what could be more important? Nothing was ever 'only TV'. Everything was always important on TV – more important than in life, anyway – which is why people simply could not live without it.

'Make Me a Star!'s initial ratings are good – very good. At one point they even equal the UK ratings of X-TV's 'The Island'.

And then it happens.

It all starts with a little boy called Connor, blind since birth and for the last two years in a wheelchair, as the terminal illness that is gnawing and nibbling at his brittle little bones, and wasting away his dying flesh, crawls in for the kill.

'Bright Eyes' is the song he sings. Badly. His voice is weak and thin as his body, his ill-looking face looks ghostly white as old bones in the spotlights. The audience loves him. But Connor has been marked down in pre-production as one to chuck out in the first round. The producers now have so many crippled and blind terminally ill contestants that they can't possibly use them all – so, against all rules of television ratings-chasing, they're going to have to dump a few kiddie cripples in the early rounds.

Peter Baztanza said it would be ground-breaking, cutting-edge TV that had never been done before, so its shock value would make headlines – and he was right. But for all the wrong reasons.

Connor sits there angelic in his wheelchair, oxygen tubes up his nostrils, cancer gnawing at his marrow. And every single member of the live audience loves him, pities him, cries for him, wishing him to be a winner, a star, a finalist at least, even though he can't sing a note in tune.

But there is just something about little Connor – just some indescribable X factor – that makes the audience want to cheer, love, defend and protect him.

Then the judges give their verdict:

'All I can say, young man, is that your cancer must've spread to your vocal cords,' sneers Silas Snide.

'If cancer makes people sing like that, then I'm right to be so frightened of it, so I am,' smirks Declan O'Dunk.

'Sorry, Connor, but it's a no from me,' sympathises Javina. 'Maybe you can take up another hobby apart from singing. Painting perhaps – lots of disabled people make lovely Christmas cards, don't they? And they don't have any arms or legs at all – so you're lucky really, compared to them.'

And then something strange happens: the audience boos. But they don't boo the act on stage as is usually the case.

No – they boo the judges. On live TV.

Obscenities are yelled from the audience: 'fucking disgrace'; 'you should be ashamed'; 'Silas is a wanker!'

Silas is not sure he has heard correctly. He has only ever been used to acts on stage being booed like this – not a judge, not him. Ever.

Declan and Javina look round at the audience, faces frozen in baffled confusion.

Silas Snide just keeps calm and carries on:

'So it's three no's, Connor. Good luck with the cancer. Bye-bye.'

But Connor does not go anywhere. Tracey Turntable beckons him to come off from the side of the stage – his wheelchair is electric after all, so all the kid has to do is press a button – but it is no good. He just sits there, a small silvery tear twinkling in the spotlights as it trickles down his ghost-white cheek, a droplet of the purest TV gold, as his O of an angelic mouth opens once again for an encore and his fragile, brave voice sings unaccompanied the first line of the first verse of 'Bright Eyes' into the outraged auditorium.

The audience rise as one to their feet, cheering and clapping and crying – while at the same time booing the judges. The schizophrenic tendency of a television audience to be cruel and sentimental by turns is on shocking display.

After the first chorus, Tracey Turntable walks onto the stage to wheel Connor off, and is roundly jeered when she does so. She has never been booed and heckled in her life before, has never known what the contestants go through, and when she looks out at the jeering audience feels a fear and a horror that a life of TV presenting has not prepared her for. Overcome, she bursts into tears.

'Sorry, I'm so sorry,' she wails, though the audience's boos are too loud for anyone to hear her words. Tracey runs off the stage, her hands over her ears, leaving Connor stranded and at an angle one third of the way across it.

The audience cheers, but then an assistant producer runs on-stage and pushes the wheelchair into the wings.

'Fuuuuuuuucking hell!' says Minty to her television at home, blinking at the horror before her: the turning of an audience against television itself.

The rest of the show is accompanied by the restless booing and jeering of an audience who are now wholeheartedly against the judges in everything. So much so, in fact, that Silas, Javina and Declan have to be given a police escort out of the building.

Minty knows it looks bad. Attacking – if only verbally – as well as humiliating and mocking a seven-year-old cancer-sufferer in a wheelchair is not really something that could ever look good. Peter Baztanza said it would be 'ironic' and 'post-modern'. But then, he would, wouldn't he?

She turns over to X-TV. 'The Island' is in full swing with

the audience vote expected within the hour to see which two contestants will be undergoing the first trial. The picture flicks between Danny Mambo with the contestants in the camp, and Alicia and Calvin on the yacht.

Minty knows which show looks better and which will win the most ratings. There is only one option left to her now. She picks up the phone.

*

Power destroys those who don't have it.

Truth is only what is believed.

Fame is a thing no sane man ought to make necessary to his happiness. It is but a temporary anomaly, good or bad.

That is all.

I am Rasmus, and this is my reality.

*

It's All a Load of Shit in the End

'Are you, um, sure it'll, um, work?' says Hugo.

'Oh aye,' says Colorado Colwyn, 'all you need is some wotcha-call – bird doo-dah, like – an' you-er plot down by yerr'll be producin' phenomenal lush veg.'

'Um, but...'

'Now then,' continues Colwyn, 'as I says, the old boy I knows, he've got all these pigeons, see, what makes a lot of...'

'Um, poo?'

'*Hexactly*,' says Colwyn.

'White gold, it is. An' this is special racin' pigeon poo, mind, so it makes things grow even quicker. I swears by it, I do.'

'Oh, um, yes, I hadn't thought of that.'

'People don't, see – they don't think bout how stuff grows, like all that there...'

Hugo follows the trajectory of Colwyn's nod to the large patch of Japanese knotweed smothering the garden's borders.

'Poor man's rhubarb, we calls it – them stems is lovely boiled up. Not now mind, but in the Springtime. I loves it!'

The dim lingering of a headache is hatching at the top of Hugo's spine.

'Now I knows what you is about to ask,' Colwyn says, 'how much this white gold is gonna be? An' I 'as to be honest with you, it's gonna cost more than your bog standard pigeon doo-dah, so don't get *hysteristical* at the price. But I thinks I can get it for you for around...about...two thousand pound.'

'Isn't that, um, rather a lot?' says Hugo, unsure if he still has that much left in his bank account.

'Oh no,' says Colwyn, 'that's the discounted price, to be honest, but it's free delivery. An old boy I know can borrow me a van an' I can be back 'ere later this afternoon. How's that?'

'Yes, um, and I'll go to the bank and, um, withdraw the, um, two thousand.'

'Plus an 'undred, like, for the van. Fair play too. Fab'lous.'

*

'That's yer lot,' says Colorado Colwyn, dumping the last of twenty sacks of dried pigeon poo at the back of the cottage. 'I's got to go now, so *hunfortunately* I's not gonna be able to 'elp you dig it in, like.'

Colwyn's fingers flick through the thick wad of twenty pound notes Hugo has handed him.

'Maybe, um, you can, um, come over tomorrow?'

'Busy now all this week, now, I is – Apache spirit's tellin' me to go up to Swansea. No idea why. But you's got to do it today – spread it all over, like.'

'What, um, all of it?'

'Oh aye. S'easy as fallin' off a log, an' I really wants to 'elp, I does, but I's just got to go now in a minute – them all-seeing spirits, they is callin' me, see.'

Colorado Colwyn looks up dutifully to the ominous-looking sky, full to bursting with angry and vengeful Apache gods.

'I's gonna be away for a while – but I's back in a few weeks, like, and by then you's gonna 'ave phenomenal veg growin' 'ere. Absolutely phenomenal, like.'

*

Later that day, Hugo opens one of the sacks. A puff of white dust mushroom-clouds out into his face as he does so, making him splutter and cough.

He lugs the bag to the middle of the vegetable plot and pours the dusty shitty contents onto the earth. As he does so, Hugo tries to imagine the barren and desolate plot bursting into life with juicy, nutritious and delicious fruitfulness the following Spring. He sees before him, in his hopeful mind's eye, a patch of land crammed with crisp carrots, cabbages, tomatoes, onions, garlic, potatoes, strawberries, raspberries and more, all succulent and life-affirming and nourished to prize-winning proportions by the two thousand pounds' worth of pigeon shit he is now raking into the stubborn soil. He does so surrounded by what looks like smoke, but is in fact a cloud of fine dust from the dried pigeon droppings, which Hugo's lungs are sucking in and out and in and out with every thirsty mouthful of air.

What with the sight gone in one eye and partial sight in the other, Hugo has not noticed the small labels on each of the sacks:

'Warning!' – the labels say – *'Do not inhale!'*

But Hugo is not thinking about his health or even his wealth. No, he is happily thinking about all the varieties of vegetables he will grow in the future which, for the first time in goodness knows how long, he is facing with an attitude of optimism and hope, rather than his usual overwhelming sense of perpetual pessimism and despair.

He digs in the guano cheerfully, and it billows around him like a cloud of radioactive fallout dust, covering everything in a fine ashen layer of dry, powdery bird poo.

Death of a Princess

Minty Chisum sits back and stares at the dark mirror of her television screen, then clicks it alive with the remote. 'The Island' is compulsive viewing, even for her.

She just knows the X-TV ratings will be huge. That is what she wanted so much for the BBC shows she commissioned from TV supremo Peter Baztanza. But instead of glory and ratings winners, the BBC is now facing disaster. She will just have to find another way to destroy Rasmus and X-TV – and she knows just the thing.

*

In the island camp, the mood is glum. The contestants have been there for almost a week now, and already four of their number are dead.

In some ways, it is an honour for anyone when any of them get voted for a challenge by the viewing public – the millions of people out there who are watching their every move 24/7. It is, indeed, a mark of popularity for viewers to vote for a contestant to undergo an experience that may well lead to a horrible and painful death.

Alberto, the Spaniard, winner of series nine of 'The House', is the first to die. He is suspended over a pond filled with ferociously ravenous piranha, when he loses his grip, slips and falls.

'Fookin' 'ell,' says Ali, his competitor in the challenge, watching Alberto plummet into the water.

But nothing happens – at first. He just bobs there like a turd, doing doggy paddle in circles. Then, amazingly, a nervous twitching of a smile appears on his frightened face as it occurs to him that this might not be for real after all – that it might just be a rehearsal, and maybe he'll be given a second chance.

He is wrong – there are no second chances on 'The Island'. Suddenly, the water is boiling and bubbling with hungry piranha, feasting on TV celebrity flesh.

Ali gulps down her fear and continues to collect the golden plastic stars suspended from the trees as screams burst from the bubbling red water below. Then, there is no noise.

Down below, inside the water, nearly not visible in its darkness, Alberto's skull grins up at Ali as if imploring her to fall and discover death for herself.

The ratings for 'The Island' that night are the biggest in TV history.

Next up is Elspeth, who Sanjay leaves behind in his quest for golden plastic stars after he hears her fall into a hidden pit. Her screaming doesn't last long – not after she's been bitten by dozens of the multicoloured venomous snakes.

Then, Dominique discovers that the indigenous peoples of the Pacific are not as friendly and harmless as her Master's degree dissertation on social anthropology might have suggested. She shares her 'death-day' with Mickey, who wishes he were back watching telly in Dudley rather than in Papua New Guinea being clubbed to death by a fearsomely black-skinned savage with a bone through his nose. Both Dominique and Mickey would no doubt be proud to know that their shrunken heads will later take pride of place outside the hut of the headman of the village. It's fame of a sort, anyway…

The headhunting trip was a special one. All four contestants – Dominique, Mickey, Dwaine, and Sade – travelled by boat to an island a few minutes away accompanied by Calvin Snow who asked them how they were feeling, and got the predictable responses – excited, up for it, well cool, and basically at peace with the universe. But only two survivors went back to the camp with him: Dwaine and Sade. No-one spoke a word on the return journey.

Back at the camp, Sade can't stop talking – quickly and manically, babbling half-finished sentences to anyone and no-one. Delayed shock, Sanjay says. So they let her talk, and follow her instructions to rearrange the camp according to the principles of Feng Shui – the reason, so Sade claims, that both Elspeth and Dominique are now dead (her own survival she puts down to 'good karma').

Benny Bun thinks of informing her that the reason four people have died is because they're on a TV reality show where the death of most contestants is not only possible, but highly probable – which is sort of the whole point. But he lets *'Shah-day'* natter on, just because she looks well upset. Probably 'on the blob or somefink', thinks Benny Bun.

Sade breaks down, sobbing hysterically and curling into a foetus shape on the forest floor.

'Awww, there there, babes,' says Angharad, stroking Sade's hair as though she were an injured pussycat. 'Death's part of life, all part of the *amaaaaazing* balance of the universe.' It is a line she has heard on one of the vampire movie DVDs she watches all the time at home – though she has added the *amaaaaazing* bit herself, which makes it sound more, well, 'amazing', she thinks.

'I know I know,' says Sade, sitting up and wiping her eyes on camera.

'Life's a journey, so death is not the end, babes,' says Angharad. 'I read that in OK magazine, I did. Catherina Zeta Jones.'

Ali brings over Sade's bag of crystals and hands them to her.

'They was on the ground, like, so I copped a feel of 'em. Hope yez don't mind, like...'

'No, Ali, thank you.'

'Fookin' mumbo jumbo if y'asks me like...'

'Not at all,' says Sade. 'Crystal therapy has basically been scientifically proven to cleanse your *chi* and realign your *chakras* – my guru says so.'

'Funnily enough, I'm just realigning me chakras right now. Got no idea how me bollocks get so bloomin twisted dahn there!' says Benny Bun, giggling with Dwaine.

Suddenly, Tub springs out of the jungle, three dead snakes dangling from his hands. Benny and Dwaine jump up ready to defend themselves.

'Me dinner,' grunts Tub, who immediately starts gutting and skinning his quarry by his camp bed. One snake appears to be still moving.

'Golly, have I missed anything?' says Princess Sarah, returning from the jungle toilet.

'Nah,' says Benny Bun, 'you ain't missed nuffink, sweetheart.'

'Gosh, it looks rather hairy,' says Princess Sarah, thinking of the task ahead.

'Like my bollocks then,' says Benny, but Dwaine doesn't giggle this time.

That's because it's already been announced which four contestants, as selected by the public online vote, by a good number of the many millions watching 'The Island' on TV around the world, will undergo the next challenge. They are: Benny, Dwaine, Angharad and Princess Sarah.

Angharad, winner of 'The House' series five, blinks in bafflement that she has been chosen, says 'Amaaaaazing!' as per usual, and then, when the news has sunk in, wonders if she'll ever be seeing Wales and mam again.

Princess Sarah says 'Golly!' and thinks of the money, and her children, and her debts, and her family. And the money.

Dwaine, meanwhile, is silent, a sudden seriousness seizing him, as he mentally prepares himself to survive the coming ordeal – or, if necessary, to die like a man.

It's all captured on camera – every moment of their lives in the camp and every trial the contestants are put through – and will be there, forever, on the Internet too, a testament to lives lived and deaths died.

*

The four contestants sit in their harnesses and pull themselves over a small lake in which lots of little logs seem to be floating. But these are not little logs. No, they are in fact great big crocodiles, with big jaws and big teeth, and big appetites too, especially as they have been deliberately starved in cages for weeks by the production crew (who have researched thoroughly the techniques used in The Colosseum of Ancient Rome).

Danny Mambo grins as the contestants are winched into place.

'Good luck guys, like,' says Calvin on the tannoy.

Some mumbled *thank you's* emerge from several dry throats with regurgitated dribbles of nervous acidic sick.

'Get as many stars as possible guys to get some extra food for later! Bring it on!' blares Alicia's tannoyed tones.

'Readeeeee, steadeeeee...' says Danny Mambo, watching the contestants hanging high above the water, and then:

'Go go go go go!'

He punches the air to start the challenge.

The four contestants in their harnesses start to pull themselves along the wires high above the lake. Crocodilian tails swish in the water below.

Benny Bun gets the first star, then Dwaine another.

Angharad dangles in her harness trying to do the 'visualisation' exercises Sade has taught her, but all she can see is dinosaur things splashing beneath her, their jaws open enough for her to see the rows and rows of amazingly triangular teeth they have.

Princess Sarah sees that the boys have already grabbed two stars.

'Come on, chapesses,' she shouts. 'Don't let's let the boys win!'

'Yay! Go girl!' yells Alicia on the tannoy as Danny and the crew watch the princess leaning out of her harness to try and grab the golden plastic star which will mean more food to cook that evening.

And then, she falls.

'Oh golly,' thinks Princess Sarah, hitting the water. Within two seconds her royal skull is crunched to a bloody pulp in a crocodile's massive powerful jaws like a giant gingery malteser.

Angharad follows Princess Sarah within the minute, trembling so much after seeing what happened to her team-mate that she actually tips herself from her harness into the deadly depths below.

'Amaaaaazing!' she screams as she falls – which, as last words go, is pretty brave, many viewers think.

Her last thought is not of Wales or mam as she expected, but of Sir Anthony Hopkins eating a tin of pineapple chunks. She does not live long enough to wonder why.

The crocodiles think that, all in all, they prefer jungle pig to the bony apes that keep falling into the water, though they're so hungry that these skinny TV celebrity monkeys will have to do for now.

Island Ratings

The ratings for 'The Island' that evening break all records.

There are predictable protests, of course – though not from royalists, most of whom despised the debt-ridden embarrassment called Princess Sarah, but from those old-fashioned luddites and traditionalists who insist on thinking that killing people as part of a reality TV show is beyond the pale.

But the death of a princess – or any of the other contestants on 'The Island' TV show – is not what is causing most criticism and consternation amongst politicians and public alike in that week's news.

That honour goes to little Connor, the terminally-ill kid wheezing in his wheelchair singing 'Bright Eyes', whose dreams the BBC judges Silas Snide, Declan O'Dunk and Javina Jelly have shattered.

'The Island' is by far the bigger ratings success, and the

producers on the yacht, as well as the presenters Alicia and Calvin, realise just how ground-breaking this all is. Rasmus Skypes them all personally to thank them for their hard work, and an email confirms that they are to be awarded another pay rise and a large bonus.

*

Coincidentally, at the exact same time, a rat is gnawing at X-TV producer Debbie Owen-Tudor's eyeball in the middle of a Swiss forest overlooking a lake. It is where she has gone to take the bottle of sleeping pills she persuaded her mother to post her at the psychiatric ward after the clinic explained that they wouldn't be able to euthanize her until the following year due to an enormous backlog of suicidalists.

Her last thoughts are not of babies or motherhood, as she expected, but of being stung by a bee at a birthday party in her garden when she was little and how her mother hugged her better. When she remembers that, she does not want to die any more. She wants to go home to see her mother, hug her and tell her how much she loves her. But it is too late. She passes out and is soon dead. The rats in Switzerland have a feast that evening, and for the next two days too, until Debs' body is found, eyeless and gnawed raw in a pretty lakeside grove.

Arrest

Minty grins. Her first smile for quite some time.

This is more like it!

She turns the volume up on the TV news:

'Rasmus Karn is a fraudster who faked ratings, rigged votes and staged pre-rehearsed reality TV shows – that's the incredible accusation being made tonight following the arrest of the owner of X-TV.'

Harry Hussain, vibrant and diverse TV reporter, stares into the TV camera and the nation's living rooms as he stands outside a West London police station getting ever wetter in the drizzle. He can't use an umbrella because then he wouldn't be able to wave his hands and arms around just like they taught him at TV training college, and how then could he express the incredible urgency of this incredible news story to his audience?

'It certainly is an incredible accusation, Harry,' says Jules Hymen, forty-something newsreader, in the studio. 'Do we know if the police are going to make another announcement this evening?'

Harry now has several school kids making rude gestures behind his head.

'No,' he nods, 'not as yet, Jules, though it really was an incredible announcement earlier today which stated that Rasmus Karn, who some call the saviour of reality TV, had been arrested.'

'So what can we expect to happen now, Harry?'

'Well, Jules, the police now incredibly only have forty-eight hours to question Mr Karn, by which time, at the end of the day, they have to charge him.'

'And if they don't charge him by the end of today?'

'Well, Jules, at the end of the day, which is not the end of today, but which is at the end of the day the day after tomorrow, the police will, incredibly, have to release him.'

'Incredible. So he could be freed at any time?'

'Yes, Jules, that is indeed incredibly the case.'

Sad newsreader faces raise eyebrows and nod their concern into the camera – Harry to Jules, Jules to Harry, and Harry and Jules together to the viewers inside the eye of the TV cameras.

The terrestrial TV channel owners have quietly ordered that their presenters show no mercy to Rasmus – treat him as guilty until proven innocent. This may or may not be connected to the fact that their ratings have halved since X-TV started broadcasting. The BBC and other channels are reporting the story in a similar manner, while all the time insisting that their news reporting is entirely impartial.

Minty watches all the TV news reports transfixed. She is gleeful at her contribution to creating the headlines that day, and is enjoying a rare success amongst all the failures.

'Make Me a Star!' has been an unsalvageable disaster, and the BBC's other shows are all sinking beneath X-TV's overwhelming and oppressive ratings success.

So this, Minty convinces herself, is the only way.

When your back's against the wall and you're really up against it, then you have to use any and every weapon at your disposal to try and win the day.

All's fair in love and war - and television.

In and Out

Rasmus was arrested on the doorstep of his Holland Park mansion earlier that day, and taken, together with laptop, in a van to the police station.

Then, as now, he had a look of total equanimity on his face, as if he knows something - *as if he just knows.*

It is this that so disconcerts all the police officers. They know some suspects look sad and forlorn, others get arrogant and cocky, some get violent and/or try to scarper, and yet others grin like loons - especially as so many suspects they arrest are drunk or stoned or well mental anyway. But they have never held a suspect who looks as calm as Rasmus, as confident in his demeanour, as almost-saintly. It is unnerving.

Rasmus tells them the truth when he is questioned. He did not fake ratings, he did not rig votes, he did not stage reality TV shows - though, of course, all shows are planned and rehearsed - and he is not guilty of fraud.

His interrogators declare that they have evidence to the

contrary. Rasmus does not seem surprised. This makes him guilty in their eyes.

Eventually, he is bailed, pending enquiries. The police tell him to stay at his London address.

He is released onto the street, where the press is a frenzied mob, shouting and yelling and screaming as Rasmus exits protected by Thursday and other body guards.

Paparazzi cameras flash and whirr. TV cameras capture the moment and relay it live to viewers at home. Most have never before seen the man who owns the TV channels they watch every day, and many have no idea what he looks like, even in our visuate age.

But immediately they can see, even via a screen, that the man carries with him a certain aura, a strange indescribable 'other-world' quality, something that seems to mark him out as being different from those around him.

Minty's face crawls and squirms with spite as she watches this on television, a rare and genuine grin signifying that, at last, she is winning - beating them all at their own game. Beating X-TV. Beating Rasmus Karn.

'I come Thursday,' says a voice from the hall. It is Oksana, the cleaner.

'*Tuesday*,' shouts Minty. 'Now fuck off.'

'You go to the hell,' Oksana whispers as she leaves.

One day, she thinks, she will get back at this horrible woman with the silly red hair who says such terrible things.

Minty does not care about her cleaner, as she does not care about her staff, or any of the people hurt on reality TV shows. She thinks about Princess Sarah, and just how great it looked when those crocodile dinosaur jaws clamped down on her stupid and over-privileged ginger royal head.

'You got what you fucking wanted, bitch, so what's the problem?' Minty yells at yet another pretty soft-focused photo of the princess on the TV screen in the next news feature story.

'You got your fame, which is what they all always want on TV, and you got your money too, after pissing away all your –

AKA *our* – cash on playing royal fucking fairytales. What the fuck else did you expect, cunt?!'

Minty flicks through the channels until she gets to X-TV again.

On 'The Island', the contestants are awaiting news of who will be doing the next challenge.

Minty has to admit, despite herself, that 'The Island' is the best – and most addictive – thing on TV at the moment by a mile.

*

A gem cannot be polished without frustration, nor a man perfected without trials.

Now the end-game has begun, the process has started which will lead to the inevitable conclusion so long expected.

Time is all there is, and all there is to do is wait. It will be a double pleasure to deceive the deceiver, when the time comes.

It is a dangerous strategy, but never was anything great achieved without danger. He who conquers himself is greater than he who takes a city.

But I can see precisely what is happening, though it may not appear apparent to most. Vision is the art of seeing things invisible – and I can see.

No terror is squeezing my heart, no fear consuming my soul. I shall merely wait. And wait. And wait.

Never confuse a single defeat with a final defeat – or a victory, for that matter.

I am Rasmus, and this is my reality.

*

Minty's Dreaming

Minty sleeps, deep in dreams.

She is back at the party at Clevedon. Lucinda is there, talking to Rasmus.

Words unclear and blurred. Can't hear what they're saying.

But don't have to hear. Know what they're saying.

Minty sent Lucinda there herself. Knows what is happening.

Now she sees Lucinda in her living room. The meeting they had.

Discussions. Offers. Plans.

Calvin Snow. At the party. Good-looking boy. Always looks so sad.

Minty has met him.

'It were reet good at first, like,' he said. 'But then...'

Seen it all before.

Stop the world I wanna get off. But too late. Too late.

Make it stop. He wants to make it stop.

Now at the party again. Calvin talking to Gary Wu.

Then Calvin talking to a dwarf. Then Calvin talking to Lucinda.

Elvis song. Minty leaves with Toby. Calvin and Lucinda, talking.

Plan in action. Strategy fixed. Touch paper lit. No way back now.

Minty knows enough.

X-TV is finished.

Colours and sounds. Dream is beautiful.

Minty wakes up with a gasp.

Reality leaves a lot to the imagination.

London's Burning

It all starts in Pudding Lane within an hour of X-TV being taken off the air.

Not *the* Pudding Lane in The City where the Great Fire of 1666 started. No – Pudding Lane, Muswell Hill. It begins when a brick is thrown by a hand belonging to a teenaged body and brain so annoyed at the sudden silent blackness of a TV screen that they hurl it at a passing police car. Though no real damage is caused, and no-one is hurt, this leads to more police cars being called to the area, and away from other less salubrious neighbourhoods. Within an hour, a full scale disturbance is underway. Soon, it spreads form leafy Muswell Hill to the rougher areas nearby - Wood Green, Turnpike Lane, Tottenham.

This is the moment the police have been dreading - when the fizzing tension of dissatisfaction and anger finally explodes into something more, a thing they cannot control. A terrorful monster let loose - the thrust of London.

That first brick is thrown by a privately-educated nineteen-year-old film school student called Wills, a big fan of 'The Island', whose mother is a producer of TV documentaries as well as a multi-millionaire (thanks to massive house price rises in the area). The brick-thrower's absent father is a university professor, an academic expert on social inclusion and the kind of inner city deprivation neither he, nor his son, nor his ex-wife have ever experienced and never will.

After chucking that brick, Wills takes a *selfie* on his mobile phone. He is careful to get the crashed police patrol car in the background.

From Muswell Hill, Facebook/Twitter messages disseminate the news of disturbances - and so they spread, like a virus of wildfire. Violent outbreaks are soon reported in all parts of London.

In The City, within minutes, rampaging packs of rioters run wild like dogs, howling their joy at the noise of destruction. They smash and grab and loot what they can. Canary Wharf, the heart of the financial industry, is locked down to prevent rioters entering the tower and destroying everything - and perhaps everyone - inside. One mob protects itself against another in our old and cosy ochlocracy.

In the West End, Soho, Leicester Square, Piccadilly, Regent Street and Oxford Street surrender to the looters, whose choice items are designer clothes and shoes, computers and games, mobile phones and the inevitable flat-screen, wide-screen, *huge*-screen TVs.

To the east, the poorer areas all burst like nail-bombs: Hackney and Haringey, Newham and Poplar, Stratford, Leyton, Barking, Dagenham, West Ham - all going up in flames. And the rioting is spreading outwards and infecting neighbouring areas, rich or poor, into the semi-detached Metroland suburbs.

In the north, even Hendon and Hampstead succumb, and Highgate too, though the violence seems somewhat politer there - almost as if people are expecting there to be intervals.

In the south, from Deptford to Peckham to Brixton to Woolwich, to genteel and popular Wimbledon and Putney and Clapham and Bromley, chaos reigns.

And to the west, Acton and Ealing burn as brightly as Hammersmith and Fulham. Riots even break out in Chiswick, Richmond and Twickenham, as looters from poorer areas and council estates move in for the kill.

In the affluent air of leafy London lanes hangs the unmistakable pong of careworn worry. How can this be happening? And where the hell are the police?

But the deepest concern of residents revolves around that most pressing and urgent issue of all: *what will all this mean for house prices?*

Predictably, what few police there are seem hopelessly outnumbered and unprepared, and are also strangely absent from many streets. This is largely because what police there are available have been called to protect the most important areas in The West End, Kensington and Westminster - such as Downing Street, The Houses of Parliament, Buckingham Palace, the great museums and galleries, and all the BBC buildings, of course.

The British public might not realise it quite yet, but they are on their own. No police will come to save them, not for a long time, though the TV news is constantly saying that reinforcements are on their way.

But there is a limit to how many buildings can be protected. So nobody is protecting Alexandra Palace, the birthplace of television, where the BBC first broadcast from in 1936 to an affluent aristocratic audience of four thousand who watched on television sets with handkerchief-sized screens which cost, in real terms, as much as brand new mid-range car does now. Soon, the building is an inferno that can be seen from miles around. The birthplace of TV is burning, as fired up as the firestarters.

From London, the riots spread all over the country, especially when people can see on their TV screens pictures of rioters and looters in London not being stopped, or even challenged, by the police - because there are hardly any plods around anyway. They spread to Kent, Essex, Sussex, the Midlands, the North, the East, the West, Wales and Scotland - to all big cities, small cities, towns and even to some villages, including the affluent hamlets of the Home Counties.

Everywhere, the endless ephebic hordes, muscles taut and sinews primed to break the chains of obedience and enforced boredom, smash and lash and ravage throughout the unhappy land, taking what they have never been given, forcing a watching world to listen, free at last to do what they want, be who they want and take whatever or whomever they want at will. It makes great telly, which is why all TV channels are broadcasting pictures of the rioting live.

Eventually, the army arrives with armoured cars on the streets of London and other major cities. But the crowds are not intimidated by the military uniforms and guns: they seem to show no fear any more, no deference or obedience to any figure of authority. Not now their favourite TV channel has been taken away.

The rioters are the people. They are young, old, black, white, male, female, gay, straight, rich, poor, right and wrong.

And they just want to watch TV.

More specifically, they just want to watch X-TV.

How dare the powers-that-be tell them they can't and take it off the air!

It is time to fight the power.

And why on earth shouldn't they riot? thinks Minty. These kids know their generation is well and truly screwed no matter what. They know - even the better off ones - that they'll never earn as much as their parents, that they'll never be able to afford the houses they were brought up in, that they'll never have it so good. No wonder they feel angry and betrayed. Well, wouldn't you?

Similar events are taking place in other countries that have also taken X-TV off the air that night, having co-ordinated with UK authorities earlier. In France, Paris suburbs explode into riots, which start as a protest against the banning of X-TV but which soon become a left-wing revolutionary struggle against capitalism and Anglo-Saxon-style austerity cuts.

In Spain, Italy and other Catholic countries, the Pope's announcement that X-TV is the creation of the devil - (something which God apparently didn't tell him until several months after the channel started broadcasting) - leads directly to X-TV being banned. The streets erupt with running battles between the police and young people, mostly unemployed and glad to be given the chance to vent their frustration and actually *do something* for a change.

All other countries that have banned X-TV face similar disturbances - all except China, which has only ever allowed limited coverage of the channel for programmes approved by its Harmony and Happiness Central Committee.

There are reports coming in from the USA of serious violence and gun crime too, though it's hard to tell if this is any more than usual in the major cities.

But Britain leads the way, and its cities are burning tonight, beacons to boredom and a new, changed world.

Minty watches the Prime Minister's broadcast that evening.

She knows that the people do not support his decision to block X-TV. She knows, too, that this is the spark that lit the flames of the riots in the first place. But she also knows that he cannot back down, not when real, verifiable evidence has been presented that Rasmus Karn and X-TV rigged votes, faked reality TV shows and defrauded the Great British Public and the rest of the world too.

Minty grins to herself at her ingenuity and cleverness. She is finally getting the better of Rasmus and will soon defeat him, though not via the terrible trials and travails of television, but through good old-fashioned skullduggery. The money Minty has paid to those who have helped her achieve this is perhaps the best she has ever spent. And she intends to enjoy the entertaining suffering it brings to the max too.

She can hear the riots outside: the breaking glass, the sirens, the chaos coming from the centre of Hammersmith. This noise is not unusual for Friday or Saturday or even weekday nights in British town centres, but the volume of violence is louder than usual. And the sounds feel different and new, somehow - sober and sour and powerful. The sky glows orange and there is the smell of burning in the air.

Rasmus can smell it too, and hear the noise of the rioters, from the X-TV HQ in Hammersmith.

Outside, down there, in the streets, the people speak:

'Ras-mus Ras-mus Ras-mus!'

The chorus of chants from the crowd are broadcast around the world by TV crews. A police cordon protects the X-TV building from being stormed by supporters.

Alone in his office, Rasmus smiles at the Gillray prints on the walls, his face a picture of perfect equanimity.

But perhaps the most important person watching television that evening is Lucinda Lott-Owen, the desperately debt-laden BBC Head of Business Strategy. She lies on the sofa at home, piles of boxes and newly-bought clothes and shoes scattered all over the floor like the dead petals pulled off some enormous shopaholic flower.

She stares at the pictures on the TV news, a sad tableau of destruction duplicated in every town and city in the country, with tears welling in her eyes. She feels blank, as though someone has deleted part of her brain, as if the image she has of herself in her mind, that of a successful and capable woman, is just an amateurish and badly drawn etch-a-sketch cartoon that has now, in one little flick of a switch, been wiped forever from the world.

It is a lie, she knows – all fake: a falsehood, a deception, a confidence trick. And now she is going to be found out, to be shown as the person she really is. Because she is not strong. No, she is weak – stupid, lonely, unhappy and, most of all, she is unable to stop shopping online, by phone, in person, in any way she can.

Why was she so weak that she accepted Minty's offer of money to do what she did? Why? She hates herself for it. But at least she knows she has a problem and needs help. And she knows she can change the terrible thing she has done too.

Lucinda spends the evening sobbing in the darkness of her flat, the awful electric flickering blue of the television aglow on her face. She thinks what she will say when she makes the statement to the press in the morning, turns events around in her head, like the screws on some ancient torture machine or the coffin of her career.

People are dead because of these riots, she knows, and it is all because of her. She has been so weak – caused so much suffering – and has done, she knows, a truly stupid and wrong thing. But now she will come clean and tell the truth, no matter what.

Lucinda Lott-Owen has made up her mind. She will confess, and confess all. Even if it means the end of her career, bankruptcy, and perhaps even imprisonment.

Anyway, it could be worse. She is not sure exactly how at the moment, but she knows it could be worse – it always can be. If nothing else, working in television has taught her that.

Thank goodness she recorded all her conversations with Minty on her mobile. She knows that nobody would believe her otherwise.

I Confess

Early morning. Sleepless night. Brand new day.

Lucinda is feeling tired. More than tired – exhausted. More than exhausted – weary and weak – and sick of it all.

But relieved – so overwhelmingly relieved that she will soon be rid of the awful burden of her lies and deceit.

After waking, Lucinda Lott-Owen cries for what seems like hours – for her deeds and actions, for the hurt she has caused people, for the dishonesty, the deceit, the lies. She weeps for others, yes – she weeps for them all – but she weeps mostly for herself, just like everyone else.

It is finally over, she knows – as will be her career, her reputation, and so many friendships. But she is doing the right thing. She will admit she was wrong – and she will accept the consequences, whatever they may be.

She leaves the flat as it is, with all the clothes and the mess. It doesn't even feel like hers any more. She will not be coming back.

Outside, the cameras click and the reporters yap, as Lucinda stands on the pavement outside her apartment block. She called them earlier that morning, told them all to expect headline news re Rasmus Karn

TV cameras. BBC, ITV, the news channels – UK and foreign. Lucinda looks at them and almost smiles. How innocent they look, those all-seeing eyes, and how harmless – just machines, metal and plastic and glass. But their very simplicity makes them sinister, like monsters spawned in some high-tech swamp, dragged from the depths to test the mettle of Man, to both free and enslave him by allowing him to see himself in all his ugliness. And yet, Man still wants to look. People will always want to look – and that instinct feeds the eternal unchanging power of television.

'Hello,' says Lucinda Lott-Owen, as if introducing herself at a group therapy session, though no-one says 'hello' back. 'I have something very important to say. As you all know, I have made accusations against Rasmus Karn, and X-TV has been taken off the air, and we have all seen the riots which resulted. Well...'

The reporters and TV people remain rarely silent, sensing a scoop.

'Well,' says Lucinda, looking up at the wide expectant faces, just like she used to do in school assembly. 'I lied. I confess that I lied and that Rasmus Karn is innocent. He did not rig votes, fake TV shows or defraud the public.'

In an instant, the gasp of a hush explodes into an ear-splitting noise of shock and disbelief: the excited eruption of the snarling and squealing and barking of a herd of hacks smelling a story.

'Why d'you do it, Lucinda?' calls a voice.

'Are you sorry for what you did, Lucy?' shouts another.

'How d'we know you ain't lying again now then?' says a third.

'I...I...' says Lucinda, and eventually the pack of reporters and photographers *shhh* and *shoosh* each other down. 'I shall now play you a mobile phone recording in which Minty Chisum asks me to…well, here it is.'

Despite the muffled tininess of the recording, the voice of Minty Chisum is unmistakeable:

'I'll pay it all off – the fucking lot – all your debts, if you do this. Accuse Rasmus of being a fraud. I can give you the evidence and you can say you got it from your X-TV contacts. When this gets out, he'll be finished, and so

will X-TV. I'll give you a hundred grand now. Today. Anonymously, of course. The same again later.'

Minty, now in her plush office on the sixth floor of BBC TV Centre, has been alerted to what is happening by a phone call from Oliver Allcock, who thinks what's on TV is 'just incredible'.

She watches the live news report in stunned disbelief. The recording of her voice speaks at her from the television, as clear as the fear in a bad dream. She feels the blood seeping from her face into her stomach, where it sits congealing in a sick little puddle of dread. Her mouth hangs open in an ominous and silent 'O', as if the invisible ghost of some giant prehistoric chicken is laying an egg within it.

Lucinda peers out, seemingly directly at Minty, from the large TV screen on the wall:

'I...I...now realise that I was wrong - very wrong - in making such accusations against an innocent man,' says Lucinda. 'And I know that I have let so many people down, especially myself. I also want to say that my financial problems are no excuse for what I did. I realise that I need help...and I'm...just so…so sorry for everything...'

Lucinda fiddles with her mobile, pressing several buttons as she ignores the yells and shouts from the press - which conveniently means she does not have to look anyone in the eye.

'I...I...' Lucinda starts, and then realises she will have to shout to make herself heard over the media scrum. 'Some of it is just voices but there are some pictures from the mobile camera too, so you can see Minty - and me - clearly...'

Lucinda presses a final button, turns off her smartphone and puts it in her handbag. The reporters *shoosh* each other down again.

'I...I've just put the recording on YouTube and my Facebook page. I shall now go and make a statement to the police. Thank you. And sorry. I really am just so... sorry.'

A taxi takes Lucinda to the nearest police station where, as promised, she makes a full statement.

By the time police officers arrive at BBC Television Centre to question Minty Chisum, she is long gone. She knows it's over and the game is up.

*

And so, the truth.

At every word, a reputation dies. With every image, a lie unravels.

Now the world knows the truth. And now the world will at last be free in the knowing.

Never was anything great achieved without danger, without risk, without the possibility of disaster – though it is always possible to roll the dice in one's own favour, if one has the means, the information and the instinct to know how.

Now the truth is out, and Minty Chisum will learn what it is to taste an unwelcome fame. May she live all the days of her life.

I am Rasmus, and this is my reality.

*

A Wanted Woman

It is announced on the news that all restrictions have now been lifted and X-TV is back on the air.

'Cunt!' says Minty to her car radio, swerving to avoid a minibus.

The radio then replays the recording of Minty instructing Lucinda to set up Rasmus. It is being repeatedly played on every TV news bulletin too, with the amateurish images of Minty taken on Lucinda's mobile making the skullduggery seem even more underhand and sinister.

Minty turns the radio off, then presses a button to wind down the electric window. The windy noise thwacks her face and makes the tangling locks of her vermillion hair swirl and squirm like little red snakes.

How could it come to this? How? She was - she is still - the Director General of the BBC. But there she was, on the run like a common criminal.

But she knows exactly where she is going. It is a sensation that she hasn't felt for a very long time indeed.

Statement

A large crowd of press and TV reporters has gathered outside Rasmus's Holland Park home. They expect a statement.

Soon, Thursday opens the front door and steps out, followed by Rasmus, who speaks:

'I would like to thank all those who have supported me during this ordeal. I would also like to state clearly now that I forgive my accuser, who was obviously not motivated by malice, but simply allowed herself to be used and exploited. I shall now be going straight back to work to ensure the future success of X-TV. Thank you.'

Then Rasmus smiles at the cameras, the morning sun glinting like glass in his clear grey-green eyes, and he goes back inside the house with Thursday, back to normality and freedom, and the bright future that awaits.

Crash

A police patrol car spots Minty's BMW on the A2 heading towards Dover - a route that was already ancient when the Romans arrived and which became Watling Street to the Anglo-Saxons. An old escape route, for sure, and one used by many before in its long history to flee the consequences of their deeds and abandon their island home to melt gently back into the crowds of the Continent.

Sirens wail, the police pursuit begins. Minty's BMW speeds up. The chase is on.

After a few short minutes there are five police cars in pursuit. Traffic a few miles ahead has been diverted. The BMW is travelling at over 100 miles per hour.

In the end, it's the stinger that stops it – punctures the tyres, ends the chase. The impact makes the driver lose control of the car. It turns and spins then comes to rest at the foot of the supports of a concrete bridge, crumpled and broken and still.

Inside the twisted wreckage, the driver sits seat-belted and cushioned by the airbag. Dead as yesterday.

The police officers curse under their breath. This was not how they wanted the pursuit to end, and usually they wouldn't use a stinger for cars going so fast. But orders were orders – and they were from the very top too.

Worse is what the police officers see when they walk over to the wreckage. The body in the car is not, as expected, its owner – the middle-aged red-haired Director General of the BBC. No, it is a young white man, unshaven and dirty-looking, but with very white teeth. His neck is snapped like a matchstick and a knuckle of vertebra is visible through the torn skin. Toby Tickell. He couldn't let Minty down, not now at her moment of most need, no matter what. Rasmus or no Rasmus, he would always serve his mistress first.

It slowly dawns on the officers present that the suspect has used a decoy to draw the police away from wherever she is actually going.

'Clever girl,' says one PC to another, who then calls in what has happened on the police radio.

A description of Minty Chisum is sent out to all police forces, as well as ports and airports, and her photograph is on all TV news bulletins too. In them, she is described as 'dangerous'. It is a description with which Minty would never disagree.

Hugo coughs again, a deep spluttering cough, dry and wet at the same time, rattling his lungs raw in his thorax. It feels as though he has a cauliflower-shaped fungus growing in his chest, a mushroom cloud of mucus and mould bubbling up and spreading deep inside him.

It is still raining, but despite his chest infection, he has to get finished – fertilise the earth so he can plant the seeds and tubers that will bring him a glut of lush organic abundance in the bright green future.

He stabs the earth with the fork, turning over the clods of soggy soil, digging in the dusty pigeon guano as carefully as this very special white gold-dust deserves. It is when Hugo leans on his fork to gaze up at the concrete-grey sky that he sees something move out of the corner of his good eye. Something red. A fox? A ginger cat? No, it can't be – Hugo's garden is so desolate and barren that all wildlife resolutely avoids it, as if some ancient curse hangs over the place, like some huge and terrible monster, its claws ready to grab any passing animal and drag them to their doom.

Hugo turns and looks – and there, right next to his open back door, stands the unmistakeable figure of Minty Chisum, hair redder than Hugo remembered, but otherwise looking as intimidating and overbearing as ever.

'M...Minty,' wheezes Hugo. 'But, um, I don't...under…'

'Aren't you going to offer me a cup of tea then?' Minty says.

'Oh, um, yes, um, ccrrrgghhhhcuhcuh...'

Hugo breaks off to heave a cough into the cuffs of his coat.

Minty knows what to expect from the tabloid pap shots of him she saw earlier, but to see Hugo in the flesh still shocks her. He is smaller somehow, shrunken and old, less present in physical space. His body seems so thin that it wouldn't come as much of a surprise if she were suddenly able to see right through it.

The garden where he's been working looks as blasted, dead and muddy as any First World War trench.

*

'I suppose you've seen the news on TV?' says Minty, taking a mug of tea from Hugo in the house.

'Well, um, I...'

'What did I fucking expect really? Loyalty is for puppies, not people.'

'Not a good, um, thing to do, though, Minty - making accusations like that.'

'Good, bad - what the fuck does it mean any more?' says Minty. 'Is it good what Rasmus Karn is doing? People are dying on TV, in all sorts of horrible-as-fuck ways. Gladiators are killing each other, scooping each other's eyeballs out with their bare fucking hands, just for fun. And we're even selling babies on live TV, just for the fucking ratings. How exactly is that *good*, Hugo? Coz I'm fucked if I know.'

'You know that you'd be, um, doing the same at the, um, BBC, Minty, if you could.'

Minty can't deny this so doesn't. Instead she just stares at Hugo, trying to see the man she once knew in the shambling wreck before her. The skin stretched taut on his sunken cheeks is so pallid and translucent that she can see the whiteness of his skull.

Hugo coughs badly - again - then sips his tea. Minty notices his hands trembling. He looks about a hundred years old, like the father - or even the grandfather - of the man she once knew.

The house she sees around her looks just as decrepit, with cardboard boxes and junk and rubbish strewn everywhere. And it smells too - of misery and failure. Dead things.

'What, um, have we done, Minty?' says Hugo, a sigh wheezing through his lips with an ominous, muffled rattle.

Minty does not understand.

'You haven't done a fucking thing, Hugo,' says Minty, confused.

'No, I mean, um, what have we, you and me, and everyone, um, done – to television?'

Minty looks at Hugo, at his small weak eyes, one empty and dead, and the other faded to dullness, like a candle going out.

'What have we, um, done to TV, to everyone, to everything?'

Hugo remembers something he learnt at school: *There is no greater sorrow than the loss of one's own native land*. He didn't understand it at the time, but now he does – absolutely.

He coughs badly again, wheezing and spluttering as he excuses himself and goes upstairs. Minty sits and thinks.

It seems like another lifetime when she was in this part of the world, but it feels just like yesterday too. It was all so long ago.

Of course, her Welsh accent has been completely smothered, first by the elocution lessons she was sent to, and secondly by her conscious effort to erase all evidence of her roots after leaving South Wales for London aged eighteen.

But Minty Chisum was brought up less than ten miles from Llandoss, in a former mining village – and how she despised the place too. To the young Araminta – the aspirational name she later discovered her harsh Aunt Enid, not her mother, had chosen for her – it felt as though she had been allocated the wrong life, born to the wrong tribe, or even to the wrong species.

And oh how she hated the Valleys! They were so backward, so stuck in their ways, a land of deeply puritanical and stifling conformity, the very graveyard of ambition. And anybody who stayed in them would be going nowhere.

She had to get out – it was the only way. Minty wanted so much more. She was creative, intelligent and ambitious – a hard worker too. And stubborn, as her aunt always said, 'as the Devil himself'.

'You wicked child,' Aunt Enid would say, whenever

Araminta dared to disagree with her. 'Damn you for a little hussy – just like your mother. You'll end up down the docks with the wasters and jack tar darkies too, you mark my words!'

Minty last saw her mother when she was twelve years old and on a school trip. Mam was drunk, swaying and laughing, arm in arm with a black man outside a pub down by Swansea docks. The woman called out to her – slurred words stinking and inebriated in the air. Her daughter, Araminta, walked on and let them fade into the past behind her forever. Always did have a ruthless streak.

And Minty always knew she deserved better – she wanted more, much more, from life than the drudgery of chapel and hearth. She also knew that if she worked hard at school, and saved what money she could, she could get away from that awful, oppressive, noisome place and make a life for herself somewhere else, somewhere free and unstifled, where she could be who she wanted to be and make a life she wanted to live.

Her Aunt Enid cursed her on the day she walked out of the house, of course, told her how she'd regret it, how she'd end up a no good hussy just like her mother. But Minty just kept on walking, that little suitcase in her hand, and she didn't look back.

She never returned either, not until Aunt Enid died anyway, and then only because a small pang of gratitude kept pestering her conscience. No matter how much of an old hateful God-fearing harridan of a witch her aunt was, at least she made sure that Minty knew the importance of education, and even sent her to elocution classes so she could speak properly and well, and thus 'get on' in the world. Minty thanked her for that, at least.

After the funeral, which was attended by a handful of distant cousins whom Minty barely knew and didn't want to know either, Minty made her way back to London with the ashes.

Enid had requested that they be scattered in the sea off Gower, but Minty, an ambitious and busy BBC producer back then, just didn't have the time. So, after the urn had been sitting on the mantelpiece of her rented Earls Court flat for a fortnight, Minty decided that the ashes couldn't just stay there forever. So down they went, down into the endless Harpic seas of eternity with a firm flush of the toilet. Amen. Hallelujah!

Surprisingly, it turned out Aunt Enid had left everything to Minty, and, when the terraced house and its contents were sold off, she found herself in possession of a nice little cheque which she used to put a deposit on her first London home.

Hugo returns to the kitchen and sits down. Minty's mind snaps back to the present.

She *knows*.

'You've called them, haven't you?' she says, quietly.

Hugo looks at the floor, the saddest of smiles stretched on his thin lips hidden somewhere within the matted bushiness of his beard. Minty always just *knew* things – that was what made her so good at her job. She knew what people were up to, knew what they were thinking, knew that they were never to be trusted.

A siren wailing somewhere. Outside, in the wet Welsh landscape, the rain stops dead, as if in awe at the moment.

'It's, um, only TV, Minty,' Hugo says, wheezing. 'It's only TV – in the end.'

Minty stands up, a resigned smile struggling to form on her lips. There is no escape now, and she knows it. There is no point running.

'You take care of yourself, Hugo,' she says, before walking out of the kitchen.

Minty opens the front door of the cottage.

Sergeant Glascock-Jones is waiting outside. Uniform have been given orders from on high to arrest the suspect and hold her until Special Branch or MI5 or whoever it is arrives from London.

Less than two hours later, the suits arrive at the local police station in a large car with blacked-out windows and take charge of the suspect. Minty is led from the holding cell to the car, then driven away, to be flown back to London from the former RAF base at Swansea airport.

On the way there, the car passes close to the one-time mining village and the two-up two-down terraced house where Araminta Margaret Chisum was born.

A Very Peculiar Coup

A CPS announcement is made that all charges have been dropped against Rasmus Karn at almost the same time that a government statement is issued announcing the rescindment of all measures taken to block X-TV.

And a significant decision has been made as well. The BBC will now be merged with X-TV to create a new television superpower called 'BBC-XTV', though many favour shortening this to 'BBC-X'.

Oliver Allcock will be acting Director General of the BBC until the merger, when Rasmus Karn will become the head of the new organisation.

There will be a new location too at the newly refurbished Broadcasting House in Portland Place W1. The old concrete 'question mark' of Television Centre in West London will soon become luxury flats (is there any other kind?)

More importantly for the public, the licence fee will be completely abolished at last. Such is the wealth of X-TV, and the profits it generates by harvesting consumer information online, that no further public contribution will be needed.

The money behind all this may well be Chinese, but nobody seems to care, so long as they can watch what they want on free TV forever.

And, as Rasmus always says when asked: 'Where do people think all the money to do this came from in the first place?'

It could all be called a very peculiar coup. Indeed, the government looks more puppet-like by the day, their strings pulled by international business and foreign cash.

Polls taken online show that there is massive support for Rasmus. The public really do not care who is in charge of the country so long as the riots have stopped and order is restored.

The armed forces and the police are also on-side – obedient and loyal as dogs.

Everything has changed – and changed utterly. The funny thing is, no-one seems to know exactly how it all happened, or precisely *what* has happened either.

But one thing is certain: nothing will ever be the same again.

Back On The Air

'Yo! Danny Mambo in da house...an in da jungle...keeping it real.'

'Yay! Amazing! Bring it on!' yells Alicia, from the yacht moored off the island.

'I is da man, d'ya get me?! Forget da rest coz I is da best!'

Danny stands in the jungle clearing in his grass skirt and face paint (copied from a website all about Zulus) holding up what he likes to think is a sacred African spear but is in fact a piece of tourist tat mass-produced in the Far East that the production designer bought second-hand on eBay.

'It's reet good to be on't air again, like, int'it?' says Calvin, his cherubic cheeks blushing peachy and buttock-ripe in the heat.

'Yo – Calvin Snow in da house!!!'

'Hiya Danny!'

'Yo! Yo! Yo! Da man dem Calvin Snow!'

'It's grrrrrrreat to be back,' says Alicia.

'It's grrrrrrrrrrrrrrrrrrrrrrrrrrrrrrrrrrreat to be BACK IN DA HOUSE!' screams Danny Mambo, out-R-ing Alicia.

'Well fierce!' says Alicia.

'Fierce as da tiger in da jungle!' yells Danny, unaware that there are no tigers at all in that particular jungle now because they have all been killed by poachers to supply the South-East Asian traditional medicine market with tiger bone and teeth, or to make tigerskin rugs for Arabs and Russian oligarchs.

'Yay! Bring it on!' screams Alicia, Calvin smiling angelically beside her. 'And how're all you guys doing in the jungle?'

The surviving contestants sit on their jungle beds, looking a little dazed, and certainly more subdued than at the beginning of The Island. But they still manage to smile, despite the heat and the flies and the sadness at all the deaths. This is TV, and on TV you must be positive, enthusiastic and smile smile smile – that is the rule, the convention, the unwritten law of reality television.

'Great!' says Benny Bun; 'Well cool,' says Dwaine; 'Fookin' up for it,' says Alison, then 'Sorry, like' when she remembers she shouldn't swear live on air.

Tub is preparing his newly caught toads on sticks and ignores Alicia's question; he does not smile.

Sanjay simply nods, not wanting to say how he feels.

Sade says nothing but a fixed smile is twitching on her face, despite herself. She has seen everyone else die, so is now just basically trying to basically find closure and overcome her trauma by basically trying to love herself just a little bit more – and she can't do that while holding a conversation on a video link.

The contestants can see Alicia and Calvin on the monitor. Meanwhile, the public at home can see the jungle clearing in the main picture, with a box at the bottom left showing Alicia and Calvin on the yacht. But what the public are not told is that during the X-TV blackout, when filming was suspended, the surviving contestants spent some time on that yacht, had hot showers, slap-up meals, clean clothes, and several nights in warm feather-bedded cabins, and got to know Alicia and Calvin a bit too – all except Tub, who refused to leave the island for the luxury yacht. It seemed, at first, that the show would be cancelled, though according to the producers they were to stay put and await further instructions.

'Fahk me!' said Benny Bun as he stepped on board. It was an opinion shared by all contestants present.

This yacht and all it represents is what they all want, why they applied for their first reality TV show, and why they, as winners of those reality shows, accepted the offer from X-TV to be on 'The Island' in the first place.

To be rich and famous. To live the dream. To really *be* someone – someone really rich and famous, somebody who matters, someone who can afford a yacht.

When the producers – Ravi, Gary Wu and Mercy – saw what was happening in London, and were told they had to pull the plug on production, they wondered if this was the end.

'Good while it lasted, innit?' said Gary Wu.

'It's not over yet, yah!' said Mercy, grinning widely as she thought of the future that awaited her in Africa – as the wife of King Balthazar Barabbas, and as a newly minted Queen, no less.

The news that 'The Island' is to recommence broadcasting is a cause of celebration for the producers and the crew, but of commiseration amongst the contestants, though they hide it well, especially on camera. They know that soon, all but three of them – the three eventual winners – will be dead.

Dwaine just knows that one of the dead will be him, and he is right – he drowns whilst attempting to swim across a river. He sort of knew that this is how his life would end. He can cope with snakes and spiders and crocodiles, and can do any athletic acrobatic feat of strength and endurance. But swimming? For a black guy like him? He doesn't want to think racist thoughts but he just can't help it. Why couldn't they give him a running challenge? He could do that. But athletics in water, like a fish? You wouldn't see Usain Bolt doing that, would you, so why do they expect him to do it? It was probably institutionally racism, when you thought about it.

Two other contestants die before 'The Island' ends. Sanjay succumbs to snakes, and Sade's self-analysing head is very basically crushed into a bite-sized chunk of fresh organic meat to satisfy a crocodile's very basic primaeval hunger, and no doubt its self-actualising ambitions too.

Benny Bun survives, but is disappointed that he comes second of the final surviving three, though at least he is ahead of that great big northern lezza weirdo Ali, who comes third. But to be beaten by the overall winner Tub – that scabby beardy toad-gobbling old pikey who not only stinks like a homeless but who also has said practically nothing at all since arriving on the island, is just 'well outa order'.

'I was robbed, innit?' says Benny to Alicia when safely back sipping champagne on the yacht. 'Like, look at me – I'm well fit, good skin, *affletic,* an' I got a great big cock an' all.'

When Benny grins at her, a hungry craving shudders through her abdomen, a primal and urgent need seeping moist through the heat between her legs. After being stuck on a yacht in the middle of nowhere for what feels like years, and surrounded by people she just doesn't fancy – the sexless Ravi and Gary Wu, and the apparently asexual cherub Calvin – Alicia jumps at the chance.

Soon, the wet sounds of human insides meeting and the muddled tongues of lust are echoing off the wooden walls of Alicia's cabin.

Neither Alicia nor Benny Bun has any idea that it is all being filmed by secret cameras for possible future broadcast and enjoyment on X-TV.

New World

Rasmus watches the screen in his office on the fourteenth floor of the X-TV building in Hammersmith. His lips smile the small smile they always have, as he looks up at the Gillray characters on the walls who have been watching it all from the very beginning.

He walks to the window and surveys the expanse of London before him. In a way, it looks the same as ever – busy, grey, with the smudge of the river's dark water running through it. But it is different now – it has changed.

Rasmus knows that it is his, this city – that he has now conquered it as others have before him. He also knows that the merger of the BBC and X-TV is more of a takeover, that he is now in charge, with the support of the politicians, the police and the people.

X-TV's international operations are proceeding as planned too, with the European, Asian and American operations all reporting success after success. The ratings for 'The Island' are stunning. But the next planned reality show will be something really special that will trump anything X-TV has done before.

With the information X-TV now holds – and is continuing to gather – about every individual viewer, its potential revenue streams are limitless. The websites, the search engines, the mobile networks, the interactive TVs – all allow X-TV to know everything about everyone – to know who, what and where they are, and how they will respond to advertising too.

X-TV's operations are taking data mining and personalisation to a whole new level. The world of the internet is a world of individuals, but it is a world of individuals who are merely satellites around the great planets of our new digital solar system – Google, Facebook, X-TV. Thus is a brave new universe created.

Thursday stands next to Rasmus – leader, conqueror, saviour. They watch the city spread out before them. Like gods.

Now is the time.

*

We have come so far, we have seen so much, and now we are here, at the dawn of a new and better age.

Vision is the art of seeing what is invisible to others. And what I saw, on the day that I became Rasmus, is what

I now see in the world: a future bright and blazing through its own creation.

The past is over. Now a better present can truly begin.

I am Rasmus, and this is my reality.

*

Ach-a-fi!

'Is you OK?' says Colorado Colwyn, his face glowing with a deep reddy-brown tan thanks to an extended holiday in southern Spain.

Hugo says nothing.

Colwyn has just found Hugo lying in the back garden, half hidden by a tangled mass of Japanese knotweed, his green wellington boots poking out of it.

'Is you sleeping, boy?' he says, moving closer to where Hugo lies. 'You looks a bit...'

Dead. He looks dead. Pale as a corpse anyway.

'Oh you's done it now, Colwyn,' he says to himself, his holiday cheerfulness evaporating like the salty Mediterranean sea from his skin.

Suicide mountain looms dark as death above Llandoss, as if it's about to collapse on the cottage there and then and crush them all to nothingness.

Colorado Colwyn leans in towards Hugo's face, which is partially obscured by the twisting tendrils of Japanese knotweed.

Wheezing. He can hear wheezing. Hugo must be alive! But just how long has he been lying here? With a little shudder of horror, Colwyn realises that the Japanese knotweed seems to have grown over Hugo, and one of its feeling tendrils has found its way into one of his nostrils and another into his ear.

'Oh *ach-a-fi!*' says Colwyn, stumbling backwards. It is the Welsh for 'disgusting'.

It reminds him of an old episode of Dr Who – Tom Baker, Sarah Jane. Television is everyone's reference for reality these days. Its milestones and memories are the religious calendar of our age.

What to do? Colwyn considers leaving him there, walking away. But who else would look after Hugo? So he decides that he'll have to call an ambulance. But how can he do that? He knows that none of these people have any understanding about how the universe works and Apache gods. They'll just blame the dried bird poo for making Hugo ill. And then who would they blame? Colorado Colwyn, that's who – and he'd probably end up going inside again – especially as he's still on probation, like, for all that business with the horses.

No, he'd have to make an anonymous phone call, then scarper quick. Hugo wouldn't know any different – so long as he got the care he needed.

And so it came to pass that Hugo is taken, within the hour, to hospital in Swansea, where he is put in an Intensive Care unit, given oxygen, and hooked up to a rack of monitors that beep and squeak and flash at him throughout the night.

It does not take long for the doctors to diagnose him as suffering from bird fancier's lung, alternatively known as 'extrinsic allergic alveolitis', or 'pneumonitis', or 'histoplasmosis', if you prefer.

The diagnosis makes no mention of the displeasure of Apache gods.

Hugo Sees the Light

Hugo lies on the hospital bed, an oxygen mask over his face.

Dreaming.

In his dream, he sees a light. Bright, white – just like they say. Though everything seems in slow-motion, and colourful, then not colourful, but white, then black, then white again in the brilliance of that bright blinding light.

And Hugo can breathe - breathe without wheezing and coughing for the first time in...

And then he sees that he is not Hugo - or not Hugo as he is now. But Hugo as he was when a boy. Maybe seven. Or younger. Very young, anyway. Happy, perhaps.

There is someone else there too. A tall man, sitting in a large chair. Though he can only see his legs and arms behind the huge newspaper he is reading. It is his father.

Hugo has something to show him: a book from school - one of his poems. He tries to speak, to say something, to get his father's attention - but can't. No words seem to come - his throat is dust-dry and parched. But he so wants father to read his poem, to see how pleased his teacher was with it.

Hugo reaches out his arm, making the nurse on duty in the ward look up at him. Patients often move in their sleep, or mumble nonsense, she knows.

The boy pulls the newspaper down so he can see his father's face. He knows this is bound to make him angry, but he has to - he just has to see him.

The nurse on duty stands up: Hugo's pulse rate has caused an alarm to beep.

Hugo the boy steps back - because the face he sees is no longer his father's face at all, the usual one, the one he remembers, the one that looks a bit like his own, but redder and stubblier and 'man-lier'.

No, the face he sees is instead nothing more than a skull - a bare stark yellow-white skull, its eye sockets staring right into little Hugo, its evil teeth grinning in anger as it leans forward towards him.

'You stupid, *stupid* boy,' says the skull, before vanishing with a high-pitched scream, leaving Hugo alone and surrounded by the last long shadows of a fading light.

And then Hugo dies.

PART THREE

THE END

And so, the future is here. Here in the now.

The past is reborn as new reality – a new world for a new people.

Rebirth. Renewal. Parthenogenesis.

Time to unravel the riddle.

All things must change. And now things have changed, and changed for ever.

The old order is gone. The world of democracy which was merely an illusion – mist on the water, a fog of self-interest and mass delusion, won through bribery and maintained through greed and corruption – is blasted and shot.

Free individuals would never choose to be thus enslaved, not when the alternative is so much more real, so much more fun.

The truism holds: the people get what the people want – and what the people want is sex, death and television.

I am Rasmus, and this is my reality.

*

And So It Begins

The world is television.

The sky flickers blue electric dark. Streets are silent empty. Blue-white bursts of lightning flash like paparazzi on the skin of the night.

A deep thud of thunder sucks the silence from the air.

It groans and roars – yet, all around, the city does not stir. It is almost 4am, and London sleeps. Its tangle of streets glows sodium orange. Glistening rain-pocked roads reflect the white light of empty offices for no-one to see. Nothing moves under the storm – no cars, no people, no life. Nothing.

Rasmus watches the world. He stands on the roof, where he stood before, so long ago that it feels like years – but this time, he stands triumphant, the city spread out in submission before him, the world changed and new.

London. It was always about London, a city built on foundations of filth and fire – a city which has changed the world. It has seen everything and nothing. It is everything and nothing – and always will be.

The river. Always and forever the river running through it – cutting through it all, bubbling under the bridges it will one day mock and destroy. Not a single body of water, but a seething mass of liquid individuals, each with a character of its own – brash and bold here, skittish and nervous there – but always the water as thick and black as old, cold blood. The river has been here longer than all of this – this little city of stone and glass – and will remain long after its buildings have been ground to sand and dust too. That will never change.

The world is television, and tonight is the night the next chapter begins. Tomorrow, the future heralds a new reality.

The eyes of Rasmus see everything, scan the city, watch and wait, as always.

The storm passes. Rain and thunder rumble low over the horizon. But now, in London, the new rising dawn glows blue in the brightening sky, as if in celebration – in triumph at a victory as famous as any other this city, and the world, have ever seen.

And above it all stands Rasmus, his face clear and calm as the wind, his all-seeing eyes smiling a new world into existence.

So let it begin.

'Death Hunt' will be the TV show to end all TV shows. It will also be the first major show on BBC-X - the new channel created from a merger of BBC and X-TV.

It is several weeks since the merger, and many changes have been made in the running of both the channel and the country.

Rasmus is Head of X-TV and BBC-X - with former Acting DG Oliver Allcock kept on as BBC Head of Operations. The post of Director General has been abolished, as has the TV licence fee - something which causes spontaneous street parties up and down the land. Since then, even the doom-mongers have been made to eat their words, as television seems to have improved markedly, even in a few short weeks.

All soap operas have been moved on to a single TV channel, thus allowing fans to watch endlessly the plots recycled from Ancient Greek dramas; their detractors can easily avoid them now. Moreover, BBC-X has announced an intention to create a new Arts Channel and a new Drama Channel - the former showing theatre productions, as used to happen in the BBC's past, and the latter which will get funding from X-TV's backers and the profits from the harvesting of consumer profile information. Even the harshest critics of X-TV have had to admit that this is more than the BBC had done for the arts for years. Sport has its own free channels too, and BBC-X can outbid any competitor for the rights to the most popular sports like football or boxing, so the public at last get the option to watch these in the comfort of their own homes.

The main BBC-X channel, meanwhile, will broadcast mainstream shows such as 'Death Hunt', whilst other X-TV channels will continue to show the usual programmes, such as the perennially popular 'Granny Gang Bang', 'Dwarf Orgy, and 'Celebrity Suck-off Specials' - which generate vast revenues. A substantial slice of X-TV's profits will be donated to charity - something which gets even their fiercest critics on their side: a small percentage of those revenues would mean hundreds of millions donated to good causes.

Television is not the only place where changes have been made. The Prime Minister remains in office, but both he and his government - and Parliament - have effectively been forced to rush through emergency legislation, permitting all necessary developments to happen. Thus, 'Death Hunt' and other shows like it have now been made a legal reality in Great Britain.

Rasmus knows he can rely on the loyalty of the army and police, and no significant public resistance is expected. The world economic crisis means that no-one is prepared to argue with anything that stands any chance whatsoever of getting the country out of its present mess. 'At least there's no rioting any more' is what most people think, and opinion polls show that the present arrangement has an overall approval rating of over 85%. It seems, indeed, that peace and harmony have broken out all over the happy land.

Of the original BBC staff, only two remain with the new Corporation: Oliver Allcock and Lucinda Lott-Owen, who has received a full pardon. She is now assistant to Rasmus Karn himself, with wide-ranging responsibilities, none of which involve business strategy or finance of any kind.

Sangeeta Sacranie-Patel, previously BBC Head of Diversity, has left the television industry completely. She thinks it wise, in the circumstances, to quit and do what her parents always wanted her to do anyway - marry her cousin, an accountant in Ealing, and have babies who, she has decided, will all become doctors or accountants or lawyers when they grow up. She will definitely steer them well clear of a career in television.

Many of the original X-TV production team have moved on too.

Ravi Govinda has accepted a promotion to become the X-TV Head of Europe, which will mean he'll still be based in

London but be in charge of all programming both in the UK and throughout Europe, including Russia.

Anita is now Head of X-TV Asia, responsible for all operations in countries there, though not including China.

Mercy - now married to President Balthazar Barabbas, King of All Africans, and soon to be Emperor too, which will make Mercy herself both Queen and Empress of Africa - heads up the continent's operations. There are already plans for Africa to produce its own local reality TV shows with a particularly 'dark continent' flavour.

Sebastian von Saxonburg remains in London in charge of all X-TV sex shows - which, he has to admit, is *the* most perfect fit.

Gary Wu - mole, snitch, traitor - has left X-TV and now sits in a prison cell no bigger or smaller than that which holds Minty Chisum.

X-TV, funded largely with Chinese cash, is now dominating the airwaves in China too, though under another name. Its planned merger with local channels means a Chinese way will be found to ensure that all appropriate X-TV shows are broadcast - or remade - in the Middle Kingdom.

The USA was always going to be a challenge, but soon new mergers will be announced which will make the resultant X-TV-controlled channels dominant in the saturated broadcasting marketplace. As always, X-TV localises each country's and region's productions, ensuring a personalisation that will appeal to every individual in the world population. Television and the Internet will become ever more merged, with information about individuals ever more accurately harvested, allowing advertising to be targeted more precisely.

In such a way, with complete domination of television and, increasingly, the Internet, X-TV has become a part of each and every human life on the planet, even in the remotest regions.

The world is television indeed.

A disgraced Minty Chisum awaits her fate in a prison cell, though she knows it is not the end – not quite yet, anyway.

She remembers how it all started – her career in television, her rise to management, her reign over the whole BBC as the first female Director General. Well, at least they couldn't take that away from her, no matter what.

She always remembers how her fight-back started. How she recruited Lucinda to help her. How she got information from Calvin Snow too. It was before Rasmus's party when Calvin had come to see her.

How young he looks, thought Minty – how innocent, how hurt, how sad. Too much like an angel, a Caravaggio cherub made flesh, as if he has simply stepped out from the warm shadow of a painting into a cold and brutal world.

Calvin Snow used one word again and again: wrong.

He said he had come to believe it was all wrong – the reality TV shows, with the violence, the sex, the horror. The death.

'I'm doing this for me mum, like,' Calvin had said, shy like a schoolboy, embarrassed by life.

And so Minty took what he gave, and what Lucinda could get too, from Rasmus, from anyone. How was Minty Chisum to know it all came from Rasmus? And how could she possibly know the evidence she was collating was all fake and fabricated to deceive and entrap her?

Minty broods in the gloom, thinking thoughts of things unspoken – wondering how exactly she ended up like this, imprisoned and alone in a world that had changed so quickly, a world with which, even she had to admit, she couldn't keep up.

What was it her depressive nervy English teacher at school used to say? 'Where there's life, there's hope.' True, perhaps, though it didn't work for the teacher, who took an overdose of sleeping pills when Minty was in the fifth form. But it could be true for Minty.

There had to be a way out of this situation, somehow - there always is, if you look hard enough.

She had life, and therefore she had hope - in theory, anyway.

Death Hunt

Thursday is getting ready, watching the line-up of contestants on a TV screen.

It is a proud day for him, for it is the first time an idea of his - one based on the old days in Africa - has been made into a TV show. And 'Death Hunt' is not just another TV show either - it will be the biggest in TV history too.

His share of the profits and royalties will enable him to buy vast tracts of his homeland, if he wants, and maybe even take over his home country itself. But for now his mind is focused on the present - on the task he has to complete to survive. It almost curls in on itself, folding and imploding like a star within his skull, taking him to 'the zone', that flowing glowing meditative state, filtering out all distraction and focusing clearly on the task ahead.

The TV screen shows profiles of the contestants - all those who will be released onto the streets of London later that morning - Minty Chisum and Gary Wu are among them. But there are also criminals there: murderers, paedophiles and terrorists, yes - but also white collar criminals, all those bankers and City traders whose greed and self-enrichment has caused so much pain to so many. They are there together with their wives who lived a life of luxury on the proceeds, and whose insatiable appetites for shopping and excess were often the motivation for their husbands' wrongdoing in the first place. They will all face their fate together.

To the vast majority, it seems fair and just and right that such people will at last get what they deserved - and not avoid all real consequences of their actions, as in the old days, pre-X-TV. People remember all too well how they hadn't been made to pay before – how they got away with it – how these criminals were, at worst, just kept in luxury in comfy high-security prisons, or perhaps just moved to another job if they had lost other people's money in the financial sector, or often they just retired with their booty stashed in a labyrinthine network of offshore companies and bank accounts. Those days were gone. Now, justice will be done – true justice, not the travesty that happened before – and the punishment will fit the crime.

But it wasn't as though these people, however much hurt and pain they had caused others, were being lined up and shot like dogs – though that, indeed, would have meant justice done in many people's eyes. No, the new system was more than fair because the criminal contestants had a chance to gain their freedom. All they had to do was escape their pursuers, to survive until the end of the game – then they would be freed, though none really deserved it.

This method of justice may well be radical and new, and people were well aware of this, but the new system seemed so much fairer than the old. And anyway, it would all be on TV, not hidden away and secret, so everybody would be able to see justice being done, including the many victims of all the crimes, which seemed the fairest thing of all. Nobody really seemed to care that this would be a break from hundreds of years of legal tradition, that the sacred cows of legality were being sacrificed to create a bright new future. It could all be said to be a new version of Trial by Ordeal anyway, and that went back centuries, to the days of Anglo-Saxons and Hywel Dda. It was progress, of a sort.

As the sun rose into the dawn on that bright November Saturday, a nation born anew prepared to watch a new kind of justice in action – one which would actually both punish the wrong-doer – fairly and justly – and entertain people at the same time.

Rasmus waits. He watches the TV screen in his office –

the same room on the fourteenth floor in Hammersmith rather than the plush new office at Broadcasting House. He is alone, and looking forward to seeing Thursday on the screen soon, starring in a show of his own creation.

The faces of the Gillray grotesques grin and gimble down from the walls, caricatures of human creatures of both then and now, naked and exposed, flayed and stripped of a civilised skin. Humanity in the raw – *Ecce Homo!*

It all feels like coming home for Rasmus – as if this was always meant to be, as if everyone was just waiting for an extraordinary individual to appear and rescue the world from its corruption and banality.

A new digital century has begun.

The future is here.

The Hunted

A blast of thumping hardcore music.

A spinning camera circling 365 degrees.

A manic cheer shrieks and screams into the microphones.

Death Hunt has begun.

'Yo, London!' yells Alicia McVicar, bouncing onto the stage with her breasts.

Screams and shrieks and whoops from the crowd acknowledge her presence. They have been here in East London for hours, queuing overnight to ensure their place at this historic event.

'Bring it on!' screams Alicia to her fans. The crowd yells her words back to her, a congregation chanting a prayer.

Danny Mambo bounds onto the stage to join her.

'Yo yo yo yo yo! Danny Mambo in da house! D'ya get me?!!!'

Cheers cheers cheers. The familiar lunatic noise of a television audience driven nuts by the presence of TV cameras.

'I am da best so forget da rest! Coz tonight, we is gonna hit you wiv a kickin show, right here on BBC-X!'

'We *so* are, Danny. Because today, in less than ten minutes, the one, the only...'

'Death Hunt!'

'...will begin. Bring it on!!!'

More noisy yelling and hollering as the lights spin and the manic music booms its thumping bassline into the stadium.

On the huge screen behind Alicia and Danny, the faces and names of all the contestants – 'The Players' on the caption – are being flicked through, page by page, person by person, on a loop. Such profile information has been on public screens all night long, with short video biographies of each contestant making sure the public get to know them before the show begins.

Amongst them are Minty Chisum – whom most of the TV audience already knows as former BBC Director General, liar, cheat, all-round bitch from hell, and convicted enemy of the people.

But many other faces are familiar – well-known criminals and those imprisoned for terrorism and horrific child abuse, plus several famous businessmen and women, bankers and traders in the financial markets, the gamblers who lost other people's money and expected no justice, no reprisals, no revenge as if by right. It is this latter group of convicts that seems the most miserable, pathetic and frightened of all, with several of them desperately pleading their innocence on their video films – offering money and gifts to anyone who can now help them – something which makes the fate that is about to befall them seem all the more right.

Saturday. 9am. An early start for a reality TV show – but this one is special. 'Death Hunt' will run live on BBC-X all day Saturday, ending at 10pm that night, when the survivors will be freed and the winners announced.

If it is a success – (and why wouldn't it be?) – then other versions and variations of the format will be shown regularly on TV, both in Britain and all over the world. It is the evolution of justice for the digital age, a new page in the chapter of law and retribution, a better way to run society.

As the hour approaches, the big TV screen behind Alicia and Danny is divided into four, showing the locations where The Players will be released. The rules have been repeated regularly since the first preliminary broadcasts earlier that week.

In addition to this, a short film has been broadcast. It features a plastic dummy and shows what would happen if a contestant were to interfere with the small metal disc inserted into the back of their neck - a coin-sized disc which contains a microchip to transmit their position both to the viewers, and to their pursuers - known as The Hunters - via GPS. The film shows that the first time the chip is interfered with the dummy receives an electric shock which knocks it off its feet - a measure designed to be fair to all contestants and avoid accidents. The second time the disc is tampered with is also the last - a small and accurate explosion from the contraption ensures that the dummy's head is blown clean off, like a cork shooting out of a bottle.

The rules of the game are listed on the huge TV screen in the east London stadium and also for the viewers at home:

1) The Players have to escape from the Hunters, who will try to delete them from the game using all available methods and weapons.

2) Any Player who manages to survive until 10pm on Saturday night will be freed – and consequently exiled to a place where they can live out the remainder of their lives in peace without fear of retribution, with all living expenses and costs paid for by the organisers.

3) Any contestant who manages to kill one of The Hunters will be freed immediately, on the same basis as Rule 2, and will be awarded a prize of £1 million.

4) Hunters will be awarded £1 million for every Player they delete. The Hunter who deletes the most Players will win an additional £100 million.

5) Any member of the public giving direct assistance to The Players or The Hunters taking part in Death Hunt will do so at their own risk, and may then be considered a valid target for either and so suffer death or injury. BBC-X and its associates bear no responsibility whatsoever for any such consequences.

6) The members of the public whose information, sent in any digital form, proves most useful to The Hunters and leads directly to deletions, will be awarded a cash prize. There are a thousand prizes available, with the top ten civilian facilitators awarded £1 million each.

7) No Player is permitted to leave the confines of London. If they do so, the tracking device will be set to delete mode, thereby terminating The Player.

8) No blame or criminal responsibility will be attributed to any Hunter or Player whose actions result in the death or injury of others.

9) The police and armed forces will monitor the activities of both Players and Hunters, and have the legal right to intervene and terminate any civilians who attempt to hamper the progress of the game.

10) The decision of the creators of Death Hunt is final, and the organisers reserve the right to amend the rules at any time.

St Paul's Cathedral, Trafalgar Square, The Albert Hall, Alexandra Palace: these are the four locations from where The Players will be released, enabling the whole of London – West, East, Central, South, and North – to be covered. The TV screen shows four large chrome and silver vehicles parked at these locations – space-age looking coaches, aluminium tubes like the bodies of aeroplanes. Each holds fifty individuals.

The countdown begins:

'Ten, nine, eight...'

The crowd takes over from Alicia and Danny.

'Seven, six, five, four...'

The presenters and the crowd chant the countdown together, joined by many of those looking at screens at home, on TVs, computers, smartphones and other digital devices, all over the world.

'Three, two, one...'

A loud buzzer is heard – this is the signal for each of the doors at the end of each bus to open automatically and for The Players to emerge.

'Go go go go go!' screams Danny Mambo, punching the air.

'Bring it on!' screams Alicia McVicar, shaking her dizzy head to a blur.

The screen shows each location – the rear doors of the coaches opening, and those inside spilling out onto the street. It is a 21st century version of D-Day, but there are no bullets being fired – not yet, anyway. The Players have one advantage at least – the Hunters will only be released an hour later at 10am, thus giving The Players a one hour head-start. To run. To hide. To stay alive.

All The Players wear the same silver jump suits. All can feel the coldness of the metal discs on the back of their necks, though they do not dare touch them. All pile out onto the streets and run as fast as possible to find a hiding place, or head for somewhere they think they'll never be found.

Minty Chisum leaves her coach at Trafalgar Square. But she does not run. Instead, she stands and watches her terrified fellow contestants scattering like panicked ants around a disturbed nest.

This is what most people are like, she knows – sheep who follow the flock, who behave exactly like others, no matter what they do. And goodness knows she has worked with enough *sheeple* in TV over the years.

But Minty is not 'most people', so she will not panic. Instead, she will think it through, logically and rationally, and defeat any enemy that the world can throw up by using her fine and fearless brain. There is always a way to escape anything, if you know how, no matter how long the odds.

Hand-held camera operators film The Players, as well as fixed cameras on all major routes and at all the best-known London landmarks. There is little danger of losing track of them, as they are being tracked using GPS, enabling mobile camera teams (and The Hunters) to eventually locate The Players and film their escape and/or their last living moments.

The vast network of CCTV in London - built up so effectively over the previous decades - will also be fully exploited. Nowhere is safe.

At St Paul's Cathedral, Gary Wu jumps with the others out of the holding bus. He knows where he will go. He's an East End lad and knows the streets - knows his old manor inside out. As Players have been granted free and unlimited access to all public transport, he makes his way to Bank station, ambling on his way as if it were the most normal thing in the world - as if he were just another commuter off to do another day's hard graft.

He is worried, yes - but he can't for the life of him see the point of panicking like the others, who are running around like headless chickens, with no plan of how to survive the coming ordeal. Gary Wu is used to utter chaos and madness - he has worked for years in the TV industry, after all - so such stress does not vex or phase him at all.

He is also a Londoner and knows exactly where he's going - knows how and where to hide in this huge labyrinth of the city.

His first thought is to go down into the sewers. But that's where Hunters will expect some Players to go and hide - as if the tracking devices would stop working down there! Also, Gary has always hated rats, which is somewhat ironic considering their importance in the TV business.

So if not the sewers, then where? Gary Wu knows just the place.

On the nation's screens, people watch in their tens of millions – and the worldwide audience is expected to reach billions later in the day. They will watch to witness the contestants running like vermin, and enjoy the moment, especially the tortured and terrified looks on the faces of all the criminals, fraudsters, terrorists and enemies of the people who will be deleted that day.

They do not have to wait long.

The Hunters

Exactly an hour later, four black armoured vehicles pull up alongside the long silver coaches from which The Players disembarked almost an hour before. Armed police guard the area, in case of sabotage or any attacks from civilians, though no trouble comes. The only disturbance is when some over-enthusiastic fans try to take selfies with The Hunters and have to be restrained.

At 10am precisely a loud siren sounds. It reminds the very oldest viewers of childhood nightmares and the screeching and wailing of air raid sirens during World War Two – though this time the enemy is not in the air, but down here on the streets. The Players deserve everything they get too. 'Death Hunt' is proving enormously popular amongst the over-eighties.

The Hunters are all dressed in black – standing robotic in darkly shiny metallic suits. Each wears a protective helmet bearing a number which also relays tracking information of their precise location. There are fifty Hunters and two hundred Players – which Benny Bun reckons works out at about three or four each, give or take, though he plans to 'total' more than that himself.

'Fahk me!' he says to himself, hardly believing he's doing this, delighted that he has been given the opportunity to take part in the TV show to end all TV shows. 'The Island' was never going to be enough, he knew that all along. Just like nothing on TV is enough – ever.

As soon as he sets foot on the road at Alexandra Palace, he starts his pursuit, scanning the skyline for possible clues as to the whereabouts of his targets. He would have preferred to hunt his prey round his old manor in Dagenham, but as a former white van man, his knowledge of London's streets extends across the city.

Maybe some of the targets have tried to go to the outskirts, zone three or four, out in Enfield and Southgate or wherever – Benny will leave that lot for the others to pick off. Instead, he will aim to delete as many Players as possible in the more compact central areas which will mean a higher hit rate. He looks at the view of Central London from Alexandra Palace and decides that that is where his quarry lies.

'Game fahkin' on!' he laughs, striding across the park like The Terminator, and thinking about his mum, the slag, watching him on telly – live!

The only thing he regrets is that he cannot choose who he can kill in this game – there are at least two dozen people in Dagenham, from teachers to coppers to old mates and old girlfriends that he'd like to delete. Who knows, maybe that could be his next TV opportunity? Game over. LOL! Ha ha ha!

On another London pavement, this time near The Albert Hall, stands another 'Island' veteran – the gruff northern bouncer, Alison.

'Stop fookin' pushin' yer twat!' she yells at another Hunter behind her as she emerges onto the street by The Albert Memorial.

When she was offered the chance to take part in 'Death Hunt', she didn't think she'd want to accept, not after 'The Island' – not after seeing what she saw, not after going through what she went through.

And yet, and yet...

Sitting in her big new house, with her massive flatscreen TV, her well expensive fitted kitchen, and a brand new car in the drive, she realised that it wasn't enough – none of it. She was rich and famous, yes – but she was also bored, missing the fix of fame that reality TV gave her. More than that, she was depressed – not just a bit down, like all them supermodels "what're always fookin' moaning bout somert or t'other", but full-on *wanting-to-hide-under-the-bedclothes-and-take-an-overdose-while-slashing-your-wrists* depressed.

And her star was already beginning to fade too, her fame's sell-by date approaching fast, a mere few weeks after 'The Island' ended.

So, even though there were risks, she decided eventually that she needed to be a part of it, this new 'Death Hunt' show – really needed another fix of fame to make her feel alive, to make her feel even half-way normal. It was as if she didn't feel she existed unless she was on TV.

There was an urge in her, an aching need, for the wonderful lovely drug of fame – and she needed ever more of it just to keep breathing, just *to be.*

The only one of 'The Island' survivors who opts not to take part in Death Hunt is the overall winner, Tub, who now lives on his own personal island in the Caribbean, and who spends most of his time living at one with Nature, slaughtering and eating as much of the local wildlife as possible. His belief is the same as it has always been: the modern world is rubbish and has nothing to offer him – except pornography, of course, which is why his modest house on the island is fully wired, with every laptop always on and his favourite porn sites accessible at the gentlest gossamer caress of a keyboard touchpad. Now all he wants is to be by himself – away from the filth of hurtful, hate-filled humanity.

'Hell is other people', he knows, so he has decided to stay in his own private heaven forever, alone.

No man is an island, they say. But *they* are wrong.

TV viewers recognise Benny and Alison immediately. Most, however, don't initially recognise the Somali security guard from 'Animal Crackers', not without his uniform. He is the Hunter pushing and shoving Alison out of the vehicle at the Albert Hall (he has no idea why these infidels let women take part in this man's business anyway). When he put his name forward for the public ballot – (he decided in the end not to go back to Somalia) – he mentioned his TV experience on the application form. And so now, as one of the lucky winners, he finds himself with a large gun in his hands going off to hunt and kill infidels, and to get money for doing it. Only in this mad *kuffir* country, he thinks – anything, but anything, is possible with these infidels.

Seeing the identities of The Hunters is something that viewers have been waiting for, wondering just who from the television cast-list will be taking part. But the biggest surprise of the day comes at St Paul's, when viewers see none other than Sir Peter Baztanza, king of reality TV, emerge from an armoured vehicle with the other Hunters, grinning from ear to ear in delight at his new role as an active participant in a television show.

'Fantastic!' he says to camera. 'I'm loving this!'

He certainly does not need the money, or the fame, and there is no good reason at all for him to be taking part in this show, other than that stubborn competitive streak that makes him unable to resist a challenge – the alluring love of the chase that flows uninhibited and electric in his televisual veins.

Of course, Sir Peter Baztanza has more reason than most to take part in 'Death Hunt'. It is, in many ways, his achievement. For did he not deceive Minty Chisum, lead the BBC team astray with unworkable ideas, thus leaving the way clear for Rasmus and X-TV – where his true loyalty always lay – to gain supremacy over the ailing Corporation, leading to its demise and subsumation into BBC-X?

But perhaps the main reason he is taking part, he knows, is boredom. It has always been the same with Baztanza, from when he was a small boy. Boredom always creeps into everything he does eventually, making him do more, and more, and more – restlessly running from that lumbering beast of tedium and monotony that has been his constant companion in life, plodding behind him like a shadow, waiting to smother him to nothing with normality and routine.

Peter Baztanza made it clear to Rasmus that he wanted to take part in 'Death Hunt', despite the dangers, even though this is one TV show that was not his idea or concept - and he was granted his wish. He knows the risks, knows that a number of Hunters are likely to die that day, and that Players will be spared for deleting them.

On the street outside St Paul's Cathedral, Baztanza sniffs the air, a predator sensing the scent of prey on the wind. And then, bounding away like a big cat, he sets off down the streets towards The East End - to Tower Bridge, Docklands, Canary Wharf, The Gherkin, The City. It is a place, Peter Baztanza has always thought, that is populated by the deeply boring and uncreative, where numbers are crunched and deals are done, all in an attempt to make money from money to the benefit of no-one but the bankers and dealers themselves. It is where the biggest crimes are committed and where the biggest criminals will hide - because Baztanza knows all about human instincts, knows how people behave when under pressure, knows that animals will always go home to die.

Back in Trafalgar Square, the fourth and final armoured vehicle stands, with its cargo of Hunters finally let loose on the world. Most scatter, spreading out in regular patterns like sound waves, off to scan and search the streets for their quarry.

But one stands in the centre of Trafalgar Square, quiet and still, watching the world around him - the other Hunters, the civilians, the cars. He stands still, with a gun in his hand, a machete hanging from his belt, and a nostalgic wistful grin on his face. It is just like the old days in Africa, thinks Thursday, and he knows that this time, just like the last time, he is going to have big fun hunting the cockroaches.

The first kill is in Knightsbridge, outside Harrods. A multi-millionaire banker who made his money from derivatives has decided that his best strategy for survival is to negotiate with a Player, to offer to pay him off, with much more than the £1 million prize his death would earn his killer. It is a logical strategy, but one that fails miserably. The banker gets deleted with a shot in the head that makes it explode like a ripe cherry tomato.

'Bring it on!' screams Alicia at the huge TV screen.

'Boom boom boom! He is da man!' yells Danny Mambo.

The crowd erupts into a frenzy of cheering and whooping as this banker becomes the first Player to be killed - and by none other than Mohammed, winner of 'Execution Night - Live', whose wealthy life in The West had offered him many luxuries and excitements, but none that can compete with this.

He accepted immediately when offered the opportunity to take part in 'Death Hunt' - the thrill of killing is something Mohammed has always loved, and he is looking forward to receiving his £1 million bonus for each kill. He also hopes to be the highest scoring Hunter and win the £100 million prize.

Money, however, is not the real reason for his participation. He is not poor, after all, even though he knows that the cash he won from 'Execution Night - Live' will run out, sooner or later - and probably sooner, seeing how much he's spending on shameless infidel whores. The fact of the matter is that he just likes killing - and is good at it too. Some people are good at cooking or carpentry, for example, some at reading and writing, and some at ending life. He just has a certain skill - a *feel* for it, that's all. It's simply a talent like any other and it's his duty to Allah to use it to its full potential.

Soon after the first kill, the first self-destruct happens - on Parliament Square, just by Big Ben. A convicted paedophile called Edwin, who had served well over a decade in prison and would soon have been eligible to apply for parole, decides that he will not give all those TV viewers out there the pleasure of watching him get tortured and played with like a cornered rat. He won't give the bastards the pleasure. It was his fault, what he had done - he knew that. But he couldn't rewind time, couldn't turn back the clock and make it all better - couldn't alter the fact that he had committed those crimes, raped those terrible, beautiful children. Edwin regretted it bitterly, cursing the compulsion that had made him that way, the God that had created the hunger in him. But he would not give them – those hypocrites out there watching it all – the pleasure of seeing him suffer any more. He would have the dignity in death that had evaded him in life.

One tug at the disc on the back of his neck gives him an electric shock that propels him onto the pavement. It is only the presence of armed police that stops some passersby from intervening – not to help the man, but to kick him to death.

Edwin gets to his feet, smiles at the sky and The London Eye, his favourite landmark, so new and fresh, so full of life. It is so clear today, like on childhood days – cold but clear, like his thoughts. He turns to look at the camera positioned a few feet away, capturing every twitch he makes, picking up each breath he breathes, right up to the end.

'It's only TV,' he says, softly, to camera.

He looks up at Big Ben. Eleven minutes past ten. As good a time to die as any. He reaches behind his head and yanks at the disc embedded in the nape of his neck for a second time.

A smaller than expected bang perhaps – more of a pop, really – leaves a body standing on Westminster Bridge, headless, with little ejaculations of blood squirting from the neck. Edwin's newly detached head, meanwhile, is sent flying up in the air, blinking its last as it sails over the balustrade, before plummeting down to the river below. It enters the dark waters with a very neat and final plop.

'Yo Yo Yo! Dat bad man is well deaded, ya get me?!' screams Danny Mambo as the trunk of the paedo's silver-suited body falls forward onto the pavement – in a way that reminds Danny of one of those statues of dictators that always get pulled over in revolutions in fucked-up shitty parts of the world.

'Bring it on!!!' screams Alicia to the crowd. It goes berserk, yelling and screeching and whooping at the wonderful sight of a famous paedophile getting his head blown off on live TV.

The viewers at home are similarly jubilant, with a good many wondering why on earth they couldn't always have got rid of criminal scumbags in this way, instead of keeping them in comfort in prison for years at taxpayers' expense.

'Death Hunt' has only just begun, but already its ratings are the highest for any daytime show in British TV history. They will climb ever higher during the day, with those for the evening coverage predicted to be the highest recorded for any TV programme, ever.

The show is being watched all over the world, (something the London Mayor hopes will boost tourism significantly), and local shows are being planned on every continent too, once people everywhere see how successful the first 'Death Hunt' is in the UK.

By midday, over sixty Players have been deleted - mostly shot to pieces by The Hunters, though three have lost their heads through choice, and four have been knocked over by the traffic - including one who hit by a bicycle courier running a red light (both courier and Player ended up being crushed under the wheels of a passing seven tonne lorry).

Seventeen civilians have died so far - mostly because they got in the way of crossfire - though two have been shot by police as they tried to intervene and attack a Player. Five people have lost their lives whilst trying to take selfies.

No member of the public, in the whole of London, has tried to help any Player or attack any Hunter. Instead, the public are fully on the side of the Hunters, with regular calls, emails and tweets from them revealing clues about Players' locations, which direction they are heading in, and sometimes even detailed information about where they are hiding.

But there is no hiding place. Not any more.

Rasmus watches 'Death Hunt' on the widescreen TV on his office wall. He misses Thursday being there, but he is not alone – and not only because of the usual Gillray grotesques observing proceedings.

For sitting on the sofa is his new assistant, Lucinda Lott-Owen, debts fully paid off, who now works for X-TV's new channel BBC-X and is in charge of Public Relations. She has known Rasmus for some time, though their relationship is strictly professional.

Others are watching that day too.

Calvin Snow wakes up at midday, naked in the arms of Sebastian. He looks over to the TV, which has been on all night, on mute, just in time to see the first Hunter killed by a Player. The Hunter is Alison, whom he remembers from 'The Island', and a Player – a convicted murderer, whose rapping name is 'Skids' but who was actually christened 'Nigel' – has ambushed her in Kensington Gardens, managed to grab her gun and immediately pump her so full of bullets that her brain has barely time to register what is happening. Skids is immediately taken away to be freed and to be given his £1 million prize money.

'Yo yo yo! Dat bruvva is well sick, d'ya get me?!' yells Danny.

'Bring it on!' screams Alicia. 'So…that Player is now completely and totally free, Isn't that just amaaaaaazing?! Bring it on!'

Calvin does not need the sound on to know what the presenters are saying – he has heard it all before.

Sebastian stirs beside him. Soon, they'll have to get up and prepare for the following day's shows – the main event that Sunday will be a replay of the highlights of 'Death Hunt', but there are also all the sex shows to go through: 'Granny Gangbang', 'Dwarf Orgy,' and a US Edition of 'Celebrity Suck-Off Special'.

He has stayed up late with Sebastian, doing what he does every day these days – taking cocaine, feeling the highest high together, then watching as his producer Sebastian goes down on him. And Calvin lets him, because he doesn't really feel it, can't really feel anything any more, because everything in his life now feels fake like TV – distant, unreal, determined by forces beyond his control. Sebastian is the only person to show Calvin any affection, so he doesn't mind that much when he does those things. Anyway, he sees far worse every day on the TV shows he presents. Drugs, sex, television. What else is there?

Sebastian wakes and looks at Calvin, the angel in his bed. He wants to believe that he brings Calvin happiness, but knows he doesn't. He has never been able to understand how one so young, so beautiful, so perfect, could be so unhappy. Calvin has beauty, he has money, he has fame – all the things that everybody, no matter who they are, craves. And now he has drugs too to make him happy whenever he chooses, and Sebastian gives him sex whenever he wants it too. Calvin's life is, therefore, utterly perfect. Isn't it?

Calvin closes his eyes.

He feels responsible for all this, knows how he tried to stop it by passing on information to Lucy, though he had no idea it was all fabricated, no idea it was he who was being played.

Calvin knows too that he could have been a Player in 'Death Hunt', like Gary Wu, and that he could be on TV being hunted right now. So why isn't he? Why has Rasmus spared him? He feels dead anyway and often wishes he was, so why isn't it him up there on the screen getting shot to pieces on live TV? Maybe his punishment is to stay alive? Maybe that's it. To stay alive and suffer for his sins. Yes, that's it. Must be. Makes sense now. He sits up on the bed and leans forward to inhale a line of coke from the bedside table.

'BOOM boom boom!' screams Danny Mambo. 'Now just one hundred Player left, d'ya get me?!'

In a small studio flat, Oksana watches a TV screen.

'Ha ha ha!' she laughs as she sees the familiar red-haired woman, whose dirt and mess she used to clean up in that house in Hammersmith. Minty is running down The Embankment away from Trafalgar Square.

'You go to the hell, you *beetch*, you not laugh so much to me now hah? Now *you go to the hell!* Ha!'

'Yo yo yo! Another one bite da dust, d'ya get me?!!!'

On the screen, another Hunter is overpowered – this time next to The Tower of London. The Player – a convicted terrorist called Khalid – rips off The Hunter's helmet and starts smashing his head to a pulp. He is delighted that he'll now get his money and his freedom, but he hasn't seen the other Hunter there, hiding by the river, who has him in his sights.

'Bring it on!!!' screams Alicia, watching Khalid's criminal head pop like a toy balloon as a high calibre round is emptied into his skull by another Hunter. Khalid's killer owes his position to a public ballot – in his everyday life he is a chartered accountant called Gavin who, to alleviate the tedium of work, has spent every free moment of the past twenty years playing violent computer games.

'Amaaaaaazing!' yells Alicia McVicar as a Hunter is followed by hand-held camera into The National Gallery on Trafalgar Square, where he simultaneously kills three Players who are hiding in the ladies' toilet – as well as five civilians who got in the way by choosing the wrong time to go for a pee.

A dark grinning mask of a face grins at viewers through the TV screen: it is Thursday, and he is on a roll. So much so that he shoots another round into the camera operator – who is a manly woman called Stevie, once director of 'The You Show' on BBC1 and grateful for any work she can get now everything has changed – or at least she was, until she felt the bullets rip into her flesh and stop her unhappy heart.

'Dat is well sick, bruv!' yells Danny Mambo, doubling up with laughter. 'But ain't nuffink in dem rule what say man dem can't murk like dat, innit?!'

'Bring it on!!!' screams Alicia. 'So now there's only eighty-three Players left.'

'Eight an' free – do you get meeeee?!'

Also watching a TV screen that evening is former Director General Benjamin Cohen-Lewis, now Professor of Media Understanding at a world-renowned London University. He will soon be made a knight of the realm, he knows, and it is also rumoured that he will be elevated to the House of Lords, where he has already decided to take the title 'Lord Cohen-Lewis of Alexandra Palace'.

'Bring it on!' screams Alicia as yet more Players are deleted.

'Dat man is da man!' yells Danny Mambo as Mohammed slaughters yet more city bankers in the streets of west London – it is these deletions that get the biggest cheers, not only from the audience in the stadium in east London, but also from the viewers at home

'Less than fifty Players left now – so Bring! It! On!!!'

'Five oh – don't you know – yo yo yo!!!'

In Llandoss, Colorado Colwyn blinks baffled at his TV screen in a darkened room, necking cans of the cheapest super-strength lager that the local offy had in stock. He likes the show, even if there is a sad lack of horses.

Colwyn wonders if that boy Hugo had anything to do with 'Death Hunt', or any of the BBC shows he has watched in the past. Strange someone like that would want to come and live in a shithole Llandoss, really, like.

He found out all about Hugo in the local paper, after his cremation – and he would have gone to the funeral too, if it weren't for the danger that suspicion would fall on him if he did. Natural causes, they said. Still, Colwyn couldn't risk it, not with the South Wales Police being what they are, and not with his record. You'd think someone like that'd get more going to his send-off, mind – more than just three people: the funeral director, a nun who just liked going to funerals, and the minister who took the service. No family members there. And no representative – or any flowers – from the BBC. What a life!

Colwyn sits and stares transfixed by the blue electric glow of his TV screen, a caveman staring into the flames of a campfire, in thrall to a great fiery power he knows he will never understand.

'Bring it on!!!' yells Alicia.
'Yo yo yo!!!' booms Danny.

BBC staff are watching too, many of them wondering who will be next, waiting for the knock on the door that will mean they will be forced to face up to whatever they have done, and end up like Minty Chisum. Some have already pre-empted any suspicion falling on them by reporting others for suspect behaviour in the past to the new guard in charge at BBC-X – the logic being that nothing proves your innocence more than making an accusation against somebody else.

'Bring it on! Yo yo yo!' scream Alicia and Danny as yet more Players are deleted, yet more bystanders blown away, yet more blood splattered onto the nation's TV screens.

Marcus Boreham-Hall watches it all from his luxury apartment in Docklands, one he has hardly left since losing his contract with the BBC, his untrimmed bushy beard now taking up even more of his face than it used to. He lies in bed, from which he emerges as rarely as possible, sipping his fifth gin and tonic of the day, and mumbling away to himself like some doorway drunk with his trademark drone – something about a 'discombobulatory juxtaposition' and a 'post-modern paradigm', though not even he is sure what any of it means any more. If it ever did mean anything at all, that is…

He stares blankly at his TV screen as another Player is deleted, someone identified in the caption on screen as one of the terrorists responsible for planning attacks on London a few years before. He hates it so much, this extreme reality TV, this opium for the visuate masses, this bread and circuses charade, no matter who is being 'deleted' – but he realises too that it is just about the most exciting, addictive and fun TV show that he has ever seen. He is, in fact, busy rehearsing in his head what to say about it all on one of his weekly arts spots on TV or radio – (he knows he will be able to find some connection with Foucault or Derrida if he tries hard enough) – when he suddenly remembers that he no longer has a weekly arts spot on TV or radio.

And then there is a frantic knock at the door.

'Yo yo yo!' screams Danny Mambo, watching – together with the live audience and the viewers at home – a Player force her way into the luxury apartments and beg Marcus Boreham-Hall for sanctuary, when he unwisely unlocks and opens his door. It is Lesley Hoppity, former Controller of BBC1 and businesswoman, who has been convicted of being an enemy of the people, and forced to be on 'Death Hunt', to be hunted like an animal in front of millions for their sick entertainment – something she absolutely deplores, even while admiring the concept and wishing she could have been involved in producing the show. But her wrongdoing is too great for redemption and forgiveness, and today justice will be done.

'Please, Marcus! Help me! Please!' she cries, pushing past him into the bedroom.

Marcus was not expecting her at all. 'Greedy Lesley' was never a friend, merely a contact – a 'television friend' to network with, someone to further his career at the BBC. How dare she expect favours now! She wouldn't help him if he were the one being chased, that's for sure.

'Get out!' Marcus screams, grabbing her and pushing her towards the front door. 'Just... fuck off! Now!'

But it is too late, for at that precise moment Sir Peter Baztanza bursts into the flat, and sprays the bedroom and its two human occupants with enough bullets to kill a herd of elephants – which, interestingly, is just what his grandfather used to do all those years ago when on safari in Kenya.

'Fantastic!' says Peter Baztanza. 'I'm loving this!'

In their respective homes, many former members of BBC staff watch and worry, wondering if they'll be next, either today or in the future, now that the world *is* the future and everything has changed – and changed utterly.

The audience for 'Death Hunt' includes anyone who is anyone in the TV industry, past and present, from those who remember the old black-and-white days at Television Centre, to those who had, in part at least, been responsible for the crude infantalisation of TV in recent years, where sex and violence had become a reliable ratings-chaser for all those involved in producing and commissioning programmes.

Rasmus may well have taken this to the extreme, but he certainly didn't start the process – this, as they all knew, had started many years before, on the day that ratings overtook quality as *the* main criterion for measuring the success of television programmes. All this – BBC-X, 'Death Hunt', everything – is just the natural conclusion of that journey.

Well, where the hell else do you think it was all going to end?

'Dat is well sick, d'ya get me!' says Danny Mambo.

'Bring it on! This is just all so amaaaaaazing!' says Alicia McVicar.

'Fahk me!' screams Benny Bun as another player's head explodes. 'This is fahkin' fantaaaaastic!'

Now there are fewer than thirty Players left, and almost forty Hunters to hunt them. The odds are not looking good – for The Players anyway.

In China, Nancy Ng studies the footage closely, picking up tips from the presenters for when she will present China's own version of 'Death Hunt' (to be called 'Harmony Through

Justice'), where criminals and enemies of the state will be hunted through Beijing, Shanghai, Hong Kong and other major cities. She is determined to become the best presenter she can be, and the most famous in all of China.

Nancy's opinion is no different from that of many of the populations of all countries who feel that the punishments inflicted on The Players - on the criminals, kiddie fiddlers, terrorists, fraudsters, scammers, and especially on the City bankers and traders - are absolutely appropriate, fair and just. The time had come for someone to do something about all the scumbags in society, to wipe the bastards off the face of the earth, to give them a taste of their own medicine and to take revenge - to get mad *and* even. And why not?

All over London, The Hunters are hunting - in Kensington and Chelsea, and Hammersmith and Fulham to the west, winkling their victims out from where they are skulking under bridges and in the undergrowth of parks; in Highgate and Hampstead, Haringey and Camden to the north; in The West End, in Regent Street, St James's, Oxford Street, Mayfair; and to the east, in the original lands of London, in Stratford, Hackney, Canary Wharf and West Ham and as far as Dagenham, where Benny Bun grew up, dreaming of being famous - a wish that, thanks to the magic and wonder of television, has now come true.

'Bring it on!!!' screams Alicia McVicar.

'Yo yo yo! D'ya get me?!!!' yells Danny Mambo, as they watch Benny Bun hunting down his latest victim.

'Take that yer fahker!' screams Benny Bun as he deletes another Player - a merchant banker called Mayhew, spraying him with bullets so he stands pinned against a wall dancing into his bullet-ridden death.

'Fahk me!' yells Benny, as the corpse falls. He grins at the TV camera nearby, especially for his biggest fan who he knows will be watching - his proud mum, the slag.

It's his fifteenth deletion of the day.

The surviving Players are now down to single figures.

One is Gary Wu, whose decision to go east has proved successful. It takes a real London boy to know all the secret parts of his city, the ones where no tourist ever goes, the secret shameful out-of-the-way places where bodies get burned and buried by people you don't want to know, where a parallel world exists alongside the 'normal' one – a place where violence has always been the native language of choice.

As he runs, he regrets everything, each footfall felt keenly like a sock in the jaw, as he beats himself up over his actions.

'Stupid, stupid, stupid,' says the road as he runs. 'I was well *fick*, innit,' he thinks, 'to do it' – to betray the man who had given him his best chance in life – and all for just a few thousand poxy quid too. No amount of money could save him now.

But there is a chance. He knows that he has two options for survival. He'll either have to somehow last until the end of 'Death Hunt' at 10pm that night, or else delete one of The Hunters to secure his survival. He is sure he can do it – that he can last out that long – certain he can survive, as he has always survived on the streets, through times almost as hard and violent as the one he's living through now.

'Death Hunt' is everywhere – on every TV and laptop and smart phone screen. So it does not take much for Gary Wu to discover that he is one of the last Players left.

Whilst running through the streets he manages to lamp a passer-by and grab his mobile – which is all within the rules, he knows – and looks up live updates online. But he knows the phone will be traceable and so he can't keep it for long. When he's finished he lobs it into the river where it lands with a ploppy squelch in the black mud of the low-tide foreshore, sticking out of the scum-coloured stinking mud like a knife embedded in someone's head.

Gary Wu knows the place he's heading for. He has kept moving most of the day, running down alleys, round estates, under bridges, leaving a trail so confusing that anyone, even with GPS tracking, would find it hard to locate him. But now he is tired - exhausted after a day's running - and he has less than an hour to go before 10pm, when he can be declared a survivor of 'Death Hunt' and gain his freedom.

So now he hides in that secret place in North Woolwich, well away from the crowds - and, he hopes, well away from The Hunters - though he knows that the disc in the back of his neck makes his location known to all those who want to kill him. But all he has to do is survive another hour - less than that now.

He has to admit to himself, as he lies hidden, that the concept of this show is 'well cool, innit?' And he wishes, not for the first time, that he'd been the one to come up with the idea - that he'd pitched it to Rasmus himself, that he hadn't tried to be clever and make some dosh on the side by leaking secrets to the BBC. But how was he to know how it'd all turn out? How could he? How could anyone?

Only one Hunter is on the trail of Gary Wu - only one determined enough to find him in his East London hole, to grab the rat by its tail and pull it into shot of the nearest camera, which, in these dark and dangerous corners of the capital will be a security CCTV camera, for sure.

'Fantastic!' whispers Sir Peter Baztanza through his toothy grin. 'I'm loving it!'

Baztanza knows where he is going too, knows where to find his toad in the hole, and in fact knows this place better than Gary Wu himself - for the simple reason that his grandfather designed it as the place where a great deal of London's shit would end up. It is with a certain pride then that Baztanza enters the Sewage Treatment Works in North Woolwich, on the north side of the Thames, just east of the Royal Albert Dock - a place filled with the stinking filth of an entire city.

Gary senses his presence before he hears it, though in that dark quiet cavernous place even the smallest sound can echo like a hand-clap - and sure enough, he soon hears footsteps.

Peter Baztanza sniffs the air. The stench is unmistakeable,

but not particularly unpleasant. It smells of London, of life - of family history in Sir Peter's case. This is the stink that made the Baztanza family rich, that allowed Peter to grow up privileged, with any and every opportunity he wanted, all of which ultimately enabled him to start Grin TV and produce television programmes that would beam filth into every single household in the land - like a toilet in reverse really, he sometimes thinks.

His grandfather pumped shit out of people's homes, and now he and others like him pump crap right back in via the magic of television - and make millions from it too. It has a neat symmetry really, creates a rational pattern with its golden shitty ratio, that it might make even the most cynical stop and ponder just how far we have come, just how far we have evolved, in the century or more since Peter Baztanza's grandfather created this palace of shit, this temple of turds.

Gary Wu can hear him closer now, knows he has to do something to escape. The Hunter has a gun, after all - all Gary has is his wits. Whoever this geezer is, he has a weapon and GPS and is going to try as hard as he can to get the job done - to kill Gary Wu, delete him from the game and from the world. He makes his move.

Peter Baztanza spins around to where the sound is coming from, a grin almost splitting his face in two. Footsteps, definitely. One Player, just like it says on the GPS. A rat scampering from a cat.

'Fantastic!' says Peter Baztanza, loud enough for the word to echo throughout the huge and cavernous sewage works hall.

A yell immediately followed by a splash tells Peter that The Player's game is up.

He pounces like a panther to where the sound has come from, simultaneously watching the GPS tracker on the front of his helmet which tells him he is nearly upon his quarry.

'Please,' says Gary Wu. 'Mate...Geezer...'

But he knows he's for it now.

Game over. The End.

It was the cry of 'Fantastic!' that did it. Gary didn't expect to hear a Hunter call out like that. He recognises that voice, knows who it is. It made him lose his footing, slip and fall into the tank below – a swimming-pool-sized tank in whose muddy shit Gary is now stuck, like that mobile phone in the foreshore of the Thames. The shit is the consistency of mud too, or quicksand – and the more Gary struggles, the more he can feel it sucking him down.

Sir Peter Baztanza's grin sparkles like a rising star in the twilight of the room, as he watches this Player, a man now branded a traitor to the truth and beauty of reality TV, who betrayed Rasmus and X-TV like a common thief, flailing in the city's stinking filth.

The CCTV cameras in the sewage works film it all – creating a brief moment of quiet on the TV screens of the land as Gary Wu and Sir Peter Baztanza stare at each other, predator and wounded prey, waiting for the end. It is a scene which, in the soft blurred light of the huge cavernous hall, has a sublime chiaroscuro to it, a certain scatological elegance and beauty reminiscent of Carravagio or Georges de La Tour – a masterpiece of television, for sure, and worthy of an award.

Even Alicia McVicar and Danny Mambo are speechless on the stage as they watch the tableau on the huge TV screen before them – probably a first in both of their presenting careers.

Sir Peter Baztanza climbs up the metal stairs to a landing above the sewage tank. He will get a better shot from here, he knows – and, most importantly, it will make for great TV.

'Please,' says Gary Wu. 'Peter...Sir Peter...we're both TV professionals, innit? I could work for you, pay you back. Please...'

But it is no good, and Gary Wu's head explodes after being filled with several rounds of automatic fire. Bits of skull and hair and skin sit on the surface of the shitty mud like bits of broken pottery as the rest of his corpse sinks slowly deeper and deeper into the enveloping filth.

Sir Peter Baztanza turns to leave, sure that he must at least be in the running for the highest-scoring Hunter. But then, something happens – something that should not have happened, he knows: Peter Baztanza slips and falls. It must have been the blood and mud splashing up at him from Gary Wu that made the metal landing so slippery.

Down, down, down he goes – down to his shitty doom.

So now, Sir Peter Baztanza is now stuck in the same situation as Gary Wu, whose limp, lifeless body is slowly sinking beneath the muddy shit a few feet away.

What with his suit of armour, heavy boots and gun, and the fact he was high up on a landing when he fell, Peter Baztanza hit the surface of the shit-tank harder than Gary Wu did and so consequently finds himself, quite literally, up to his neck in the gloopy poo which surrounds him. He tries but cannot reach the sides, cannot even move his arms, both of which are stuck firm under the mud.

In the studio, Alicia and Danny Mambo watch the scene open-mouthed, as do members of the crowd – and the TV audience at home. This was not meant to happen. However, there is nothing in the rules to say that it shouldn't, couldn't or, indeed, wouldn't happen to any one of The Hunters – and that includes the most eminent, wealthy and famous ones, such as Peter Baztanza.

Most people would struggle, plead for help, beg anyone on the other side of the camera to help them, to save them, to be their friend. Not Peter Baztanza. His hyperactive brain worked out, within a split second of him slipping into the shit-tank, that these were to be his last moments of life, his swansong, his very final and first starring role on the reality TV he had done so much to create.

This clip, of his final moments, will be played and replayed for years – maybe, centuries – to come. Thus he will not spoil the televisual moment with hysterical begging or weeping. He'll go down with dignity, sure in the knowledge that he, and the reality TV he has created and over which he reigned supreme for so long will live on after his death, safe in the hands of Rasmus and X-TV, and now BBC-X, thanks in large part to his treachery. He will be remembered, he knows – replayed and re-edited a thousand thousand times – and in such a digital memory and final fame will rest his immortality.

'Fantastic!' says Sir Peter Baztanza, spluttering the word through a mouthful of faeces.

He leans his head back into the filth, so as to take one last look at the wonderful world he has created, to stare for one final time at the CCTV lens capturing for all eternity the death of a king.

The eye of the camera – the eye of God. Is there any difference anyway?

Baztanza's head sinks ever deeper into the human slurry, soon to be completely enveloped by the shit of a city.

'I'm loving this!' his sparkling grin says as the filth folds in to cover him.

And, with that, Sir Peter Baztanza disappears forever.

His grin is the last part of him to vanish into his gloopy grave, slowly going under the mud with a final sorry plop.

Minty Lives

There are only three Players left now – one in the north, one in the west, and Minty Chisum, who has stayed all along in the London she knows best – The West End, which was the first area she came to as a new 'refugee from Wales', all those years ago. It is the only place where she knows she has a hope of surviving.

She has run that day through the streets of Soho, past the editing and screening rooms she had visited so often, over to Piccadilly, across Leicester Square, through Covent Garden, and along The Embankment. It is there that she sees him.

Thursday – the attractive man she first saw in that photograph of Rasmus, the man who was there at the party, who changed Toby so much – the man who now stands before her dressed as a Hunter, a large loaded gun in his hands.

She comes face to face with him by Charing Cross, and knows this is the end. They meet in the orange-white glow of the street lights. Eyes look into eyes. Like minds meet.

But Thursday does not shoot her, as she expects. Instead, he smiles and lowers his gun.

Minty wonders what he is doing. It is his job to kill her and she almost tells him so, right there – she always hates it when participants in TV shows think they can disobey the rules!

Then Thursday slings the gun over his back and unsheaths his machete, its glinting silver blade still smeared wet with the blood of others.

He nods at her. Minty knows what that means, and so she runs – up the steps to Hungerford Bridge. Thursday follows behind – walking, not running, so as to prolong the pleasure of the chase.

The TV viewers are transfixed. It is five minutes to 10pm – and if Minty manages to survive until then she will be free.

Minty runs along Hungerford Bridge, the rattling of a train departing Charing Cross station on one side and the darkness of the river on the other.

Suddenly she stops. She can see that there is another Hunter on The South Bank by the Royal Festival Hall – he sees her and heads up to the bridge on the other side of the river.

She turns around. Thursday is there, walking slowly towards her, smiling wide, the machete glinting in his hand.

Minty looks down into the dark heavy water of The Thames churning below as it folds itself around the feet of the bridge, thick and slow as tar, a river of black lumpy blood beneath her, beckoning her down. How easy it would be to jump, to feel the water consume her, drag her deeper and deeper into its ancient black heart, and keep her there forever, like a secret.

'Oh fuck this for a game of cocks and cunts!' Minty thinks, mouthing the words – but she thinks of much else besides.

She thinks of Hugo, of what could have been; of how everything had all gone so wrong; of what she had done in her life and her career in television – or perhaps what she hadn't done – and more to the point, why?

And then Thursday is standing in front of her on the bridge, just a few feet away.

Minty knows it is all over.

'It's only TV,' she says.

Thursday grins, holding up the machete, ready to chop.

But before he can, Minty grabs the railing and launches herself from the bridge into the inky darkness below.

It's only TV, yes. But maybe TV is all there is?

Big Ben chimes ten times and the siren to end 'Death Hunt' sounds.

'Yo yo yo! You know what time it is, d'ya get me?!!!' says Danny Mambo.

There is one survivor standing amongst The Players, a convicted murderer who is greatly looking forward to his freedom, as well as four others who have managed to delete Hunters to earn their freedom.

Most Hunters survive, of course, and most Players do not.

Benny Bun does not realise his ambition of being the top scorer, however, and comes a disappointing second. The winner of 'Death Hunt', who has managed to delete the highest number of undesirables, is Mohammed, star of 'Execution Night – Live', who has enjoyed every second of his killing spree. He is now looking forward to buying up land in a country somewhere hot with his prize money, where he can conduct his own private versions of 'Death Hunt', *inshallah*.

The crowd yells and cheers and screams, a single mental organism moulded from humanity, going wild at the spectacle they have just seen – and already looking forward to the next reality TV show to hit their screens, whatever the hell that is.

'Bring it on!' yells Alicia, 'Amaaaaaazing!'

And it is.

*

And so the end, which is also the beginning.

Vision is the art of seeing things invisible.

But television is all we need to see, because it can see for us.

TV is all we need to be free – to live as free individuals – free to choose, free to decide our own destinies. Free to live and to die, free to love and to hate, free to entertain and to be entertained. Free to be human.

For if we believe in absurdities, we shall commit atrocities.

This is all there is and all we need to know:

TV is a monster
TV is a moron
TV is a madness
TV is us

I am Rasmus.

And this is reality.